Introduction
to Filter Theory

PRENTICE-HALL ELECTRICAL ENGINEERING SERIES

Introduction
to Filter Theory

DAVID E. JOHNSON

Electrical Engineering Department
Louisiana State University

PRENTICE-HALL, INC., *Englewood Cliffs, New Jersey*

Library of Congress Cataloging in Publication Data

JOHNSON, DAVID E
  Introduction to filter theory.

  Bibliography: p.
  Includes index.
  1. Electric filters. I. Title.
TK7872.F5J63   621.3815′32   75–28442
ISBN 0–13–483776–2

© 1976 by Prentice-Hall, Inc.
  Englewood Cliffs, New Jersey

10   9   8   7   6   5

Printed in the United States of America

PRENTICE-HALL INTERNATIONAL, INC., *London*
PRENTICE-HALL OF AUSTRALIA, PTY. LTD., *Sydney*
PRENTICE-HALL OF CANADA, LTD., *Toronto*
PRENTICE-HALL OF INDIA PRIVATE LIMITED, *New Delhi*
PRENTICE-HALL OF JAPAN, INC., *Tokyo*
PRENTICE-HALL OF SOUTHEAST ASIA (PTE.) LTD., *Singapore*

Contents

Preface

This book is intended for an introductory course in passive and active analog filter theory and design. Because, in the generalized sense, a filter is a device that transforms an input signal into a specified output signal, it is evident that filter theory is really circuit theory. Thus the book could be used in an intermediate circuit theory course following the usual sophomore sequence. Indeed, much of the material usually found in networks books, such as positive-real functions, Foster and Cauer networks, one- and two-port synthesis, etc., appears in this book. The book may be readily understood by a student who has had a first course in circuit theory and an elementary treatment of Laplace transforms.

The various elements of modern filter theory, such as the approximation problem, frequency transformations, and passive and active synthesis of transfer functions, are found scattered here and there in circuit textbooks and journals. It would seem appropriate therefore for these topics to be collected in one volume with a treatment sufficiently elementary to be accessible to undergraduate students, practicing engineers, and filter novices. This is the need that this book endeavors to fill.

The length of the book is such that it can be covered in one semester in an introductory filter or intermediate circuits course, or it may be useful as a supplement covering filter theory in a more comprehensive circuits course.

The approximation problem is treated in detail, including Butterworth, Chebyshev, inverse Chebyshev, Bessel, and elliptic filter functions. In the latter case, a recently derived method is employed that avoids the need to use or understand elliptic functions, thereby making this important filter case available to undergraduates.

The synthesis of low-pass, high-pass, bandpass, band-reject, all-pass, and constant-time-delay filters is considered from both the passive and active standpoints. In the passive case, lossless ladders are developed with both single and double resistance terminations. Constant-resistance networks and symmetrical lattices are also considered. The active synthesis presented uses the operational amplifier as the active element. Most of the commonly used active circuits are given and analyzed along with some new designs previously considered only in circuit theory journals.

For completeness, chapters on time-domain considerations and sensitivity are also given. In the latter case, in addition to the usual sensitivity functions, a section is devoted to variations using the shortcut concept of homogeneous functions.

The author is indebted to his students, particularly M. D. Kashefi and A. Eskandar, for their many contributions through the years to the textual material, and to his colleague, J. R. Johnson, who has been a faithful partner in developing many papers on the approximation problem. Also, particular appreciation is due colleagues J. L. Hilburn and A. H. Marshak, who supplied respectively the photographs of actual filter responses and the advice on the operational amplifier material in Chapter 10.

DAVID E. JOHNSON

Introduction
to Filter Theory

1

Introduction

1.1 HISTORICAL NOTES

An electric filter is a network that transforms an input signal in some specified way to yield a desired output signal. The signals may be considered in the time domain or in the frequency domain, and correspondingly, the output requirements of the filter may be stated in terms of time or frequency. In the latter case, a filter is often a *frequency-selective* device which passes signals of certain frequencies and blocks or attenuates signals of other frequencies.

Electric filters permeate modern technology. It is difficult to conceive of any moderately complex electronic device that does not employ a filter, in one form or another. Telephone, telegraph, television, radio, radar, sonar, and space satellites are but a few examples in the fields of modern communications and signal-processing in which filters are essential components.

Filter theory had its beginning in 1915 when Campbell in America and Wagner in Germany independently invented the electric-wave filter. The theory has evolved essentially along two independent lines, known in the literature as *classical filter theory* and *modern filter theory*. The classical theory was developed in the 1920s by Campbell, Zobel, and others, and is concerned with the design of passive lumped filters using the method of *image-parameters*. (See, for example [V], [RB]*.) This theory yields good results with a minimum of effort because of the wealth of published design information available. However, for more precise and accurate results, modern filter theory is much to be preferred.

Modern filter theory, developed in the 1930s by Cauer, Darlington, and others, is more general and more efficient than the classical theory. Essentially, it involves the approximation of the filter specifications by a transfer function, and the design of a network, using exact methods, which realizes the transfer function. Thus the problems of approximation and realization may be solved separately in an optimum and exact manner. The disadvantage of the modern theory, the complex arithmetic which it entails, has been largely overcome by the increasing availability of high-speed computers, and consequently the modern theory has become more popular than the classical theory.

Filters may be classified in a number of ways. For example, *analog* filters are used to process analog signals, that is, signals which are functions of a continuous-time variable. *Digital* filters, on the other hand, process digitized continuous waveforms. Filters may be classified as *lumped-element* or *distributed-element* devices, depending on the frequency ranges for which they are designed. Finally, we may classify filters as *passive* or *active* depending on the type elements used in their construction. Elements usually used in electric filters are resistors, capacitors, inductors, and such electron devices as transistors and operational amplifiers. In addition, some filters employ mechanical, crystal, and switching devices.

Our purpose here is to present an introduction to analog filter theory, along the so-called modern lines. We shall consider the approximation problem, the nature of the transfer functions, and both passive and active lumped-element filter realizations. A reader who is interested in more than an introduction to these topics, as well as in other topics, such as digital filters, mechanical filters, crystal filters, N-path filters, distributed-element filters, etc., is referred to such works as references [H] and [TM].

Because of the very general definition we have given of a filter, that is, a device which transforms an input to yield a specified output, filter

---

*References thus abbreviated are listed in the alphabetical sequence of the abbreviations in the Bibliography at the end of the book.

theory is actually circuit theory. Indeed, the typical intermediate book on electric circuits develops the general theory and applies it to filter design. Therefore our purpose here may be stated alternately as the presentation of an intermediate course in circuit theory with applications in modern filter theory.

1.2 PRELIMINARY DEFINITIONS

A single-input, single-output filter may be represented symbolically by Fig. 1.1, where $x(t)$ is the input signal and $y(t)$ is the output signal. If the filter is composed of linear, lumped elements, and t is a continuous-time variable, then x and y are related by a linear, ordinary, integro-differential equation, which may be Laplace-transformed, if there is no initially-stored energy, to yield

$$Y(s) = H(s)X(s) \qquad (1.1)$$

where $s = \sigma + j\omega$ is the complex frequency.

Figure 1.1. Symbolic representation of a filter.

The quantities $Y(s)$ and $X(s)$ are respectively the Laplace transforms of $y(t)$ and $x(t)$, and $H(s)$, the *network function*, is the ratio of the transformed output and input variables. When $s = j\omega$ (ω measured in rad/s), the network function is complex and may be written in the form

$$H(j\omega) = |H(j\omega)| e^{j\phi(\omega)} \qquad (1.2)$$

where $|H(j\omega)|$ is the *amplitude* or *magnitude* and $\phi(\omega)$ is the *phase*. The *amplitude* and *phase responses* are respectively the plot of $|H(j\omega)|$ and $\phi(\omega)$ versus ω, and may be used to characterize the filter. For example, if ω_1 is the frequency of a signal which is passed by the filter, then $|H(j\omega_1)|$ is relatively large, and if ω_2 is the frequency of a signal which is blocked, then $|H(j\omega_2)|$ is relatively small (and is ideally zero).

We may also express the network function in the exponential form

$$H(j\omega) = e^{-\gamma(j\omega)} \tag{1.3}$$

where

$$\gamma(j\omega) = \alpha(\omega) + j\beta(\omega) \tag{1.4}$$

In this case we have

$$\begin{aligned}\alpha(\omega) + j\beta(\omega) &= -ln\, H(j\omega)\\ &= -ln\,|H(j\omega)| - j\phi(\omega),\end{aligned}$$

or

$$\begin{aligned}\alpha(\omega) &= -ln\,|H(j\omega)|\\ \beta(\omega) &= -\phi(\omega)\end{aligned} \tag{1.5}$$

The quantities γ, α, and β are all dimensionless, α being denoted in *nepers* and β in either radians or degrees. More often, α, which is defined as *loss*, is converted to decibels (dB) by multiplying it by the factor $20 \log_{10} e$. That is, the loss in dB is given by

$$\begin{aligned}\alpha_{dB} &= (20 \log_{10} e)\alpha\\ &= -20 \log_{10} e^{-\alpha}\\ &= -20 \log_{10}|H(j\omega)|\end{aligned} \tag{1.6}$$

Another quantity of interest in filter theory is the *time delay* or *group delay* $T(\omega)$, defined by

$$T(\omega) = \frac{d\beta(\omega)}{d\omega} = -\frac{d\phi(\omega)}{d\omega}\ \text{sec} \tag{1.7}$$

The phase response and the time delay are important, as we shall see, if it is desired that the signal pass through the filter with very little distortion. In fact, as will be seen in Chapter 7, if the phase response is linear (in which case the time delay is constant), then there is absolutely no distortion in the signal.

Depending on the application of the filter, the designer may be more interested in one aspect of the filter function than another. For example, if the filter is to be frequency selective (that is, signals of some frequencies pass while those of other frequencies are blocked), the loss $\alpha(\omega)$ or, equivalently, the amplitude $|H(j\omega)|$, is the important characteristic. On the other hand, if it is of overriding importance that the output signal have a minimum of distortion, then attention should be focused on the phase response or, equivalently, the time delay.

1.3 TYPES OF FILTERS

A frequency-selective filter is one that passes signals whose frequencies are in certain ranges or bands, called the *passbands*, and blocks, or attenuates, signals whose frequencies are in other ranges, called the *stopbands*. The nature of the amplitude function $|H(j\omega)|$ or the loss function $\alpha(\omega)$ may be used to classify the various types of filters according to the location of their pass- and stopbands. An *ideal* filter is one which has a linear phase response in its passband, zero loss in its passband, and infinite loss ($|H(j\omega)| = 0$) in its stopband.

The most often encountered types of frequency-selective filters are defined as follows:

(1) A *low-pass* filter is one with a single passband between 0 and a *cutoff* frequency ω_c, with all frequencies higher than ω_c constituting the stopband. The *bandwidth* is defined as $B = \omega_c$.

(2) A *high-pass* filter is one with stopband $0 < \omega < \omega_c$, and passband $\omega > \omega_c$. (Again ω_c is defined as the cutoff frequency.)

(3) A *bandpass* filter is one with a passband between two cutoff frequencies ω_L and $\omega_U > \omega_L$ and two stopbands, $0 < \omega < \omega_L$ and $\omega > \omega_U$. The *bandwidth* is defined as $B = \omega_U - \omega_L$.

(4) A *band-reject* filter is one with a stopband $\omega_L < \omega < \omega_U$ and two passbands, $0 < \omega < \omega_L$ and $\omega > \omega_U$. (Other terms used are *band-elimination* or *band-stop*.)

(5) An *all-pass* filter is one which passes all frequencies equally well. That is, $|H(j\omega)|$ is constant for all frequencies, with the phase $\phi(\omega)$ generally a function of frequency.

Unfortunately, ideal filter behavior is physically unrealizable. This is because, as we shall see, the practical network functions $H(s)$ are ratios of polynomials and $|H(j\omega)|$ cannot possess the discontinuities necessary for the clearly-defined boundaries between the pass- and stopbands. The filter designer therefore must obtain a response which is practical and which approximates, within some specified set of tolerances, the ideal response.

As an example, the ideal amplitude response of a low-pass filter, defined by

$$|H(j\omega)| = A, \qquad 0 < \omega < \omega_c$$
$$= 0, \qquad \omega > \omega_c \tag{1.8}$$

is shown in Fig. 1.2, with a realizable approximation to the ideal represented by the solid line. The shaded areas represent a possible set of

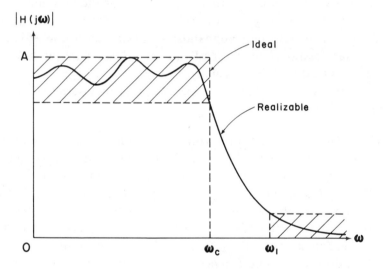

Figure 1.2. Low-pass filter responses.

tolerances within which the response must lie. The response of a low-pass filter which was actually constructed in the laboratory is shown in Fig. 1.3.

The passband $0 < \omega < \omega_c$ and the stopband $\omega > \omega_c$ are clearly indicated in the ideal case but in the realizable case the cutoff point must be defined. The usual definition is that ω_c is the point at which $|H(j\omega)|$ is $1/\sqrt{2} = 0.707$ times its maximum value, shown in Fig. 1.2 as A. Since $|H(j\omega)|^2$ is often related to power, and at ω_c, $|H(j\omega)|^2$ is half its maximum value, the point ω_c is called also the *half-power* point.

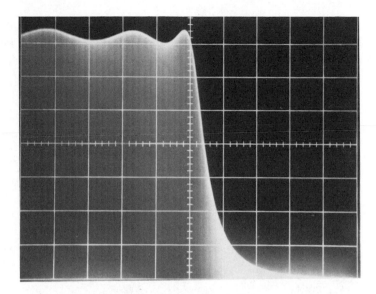

Figure 1.3. An actual low-pass filter response.

Finally, at ω_c, the loss α_{dB} is given by

$$\alpha_{dB}(\omega_c) = -20 \log_{10} \frac{A}{\sqrt{2}}$$
$$= 3 - 20 \log_{10} A$$
$$= 3 + \alpha_{dB_{min}}$$

Thus at ω_c the loss is 3 dB greater than the minimum loss $\alpha_{dB_{min}}$, and for this reason ω_c is sometimes called the 3 dB point. (It should be observed that $\log_{10} 2$ is not exactly 0.3 and thus the 3 dB point and the half-power point are not quite the same. The two terms are usually used interchangeably, however.)

It is clear from Fig. 1.2 that there is some passage in the stopband in the nonideal case (as well as some attenuation in the passband), and that it is ridiculous to say that frequencies just larger than ω_c are rejected while those just smaller are passed. Thus it is convenient to define the stopband as $\omega > \omega_1$, where as shown in Fig. 1.2, ω_1 is the point at which the response reaches and remains with increasing ω below some specified value. The interval $\omega_c < \omega < \omega_1$ in which the response is monotonically decreasing is then referred to as the *transition* band. In the case of bandpass filters, there are two transition bands, one between the passband and each of the two stopbands.

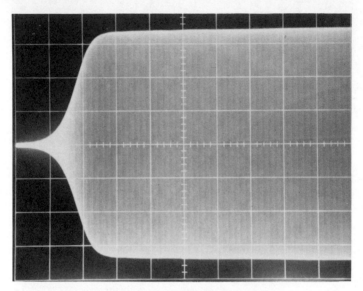

Figure 1.4. A high-pass response.

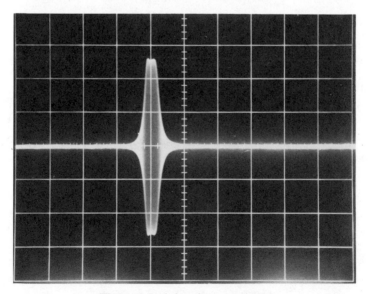

Figure 1.5. A bandpass response.

To illustrate types of filters other than low-pass, Figs. 1.4, 1.5, and 1.6 show amplitude responses of actual circuits that are, respectively, high-pass, bandpass, and band-reject filters. The responses shown are complete, rather than the top half, as was the case in Fig. 1.3.

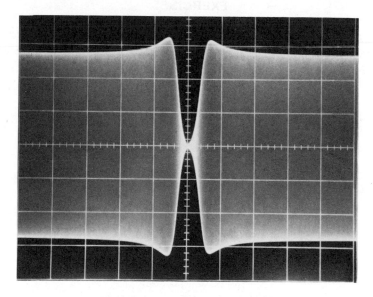

Figure 1.6. A band-reject response.

1.4 SUMMARY

This chapter introduces filter theory, discusses various types of filters such as lumped-element, passive and active analog filters, and defines the various network functions associated with filters. These functions were used to classify types of filters, such as low-pass, high-pass, band-pass, band-reject, and others, and examples were given of ideal and realizable filter responses.

In modern filter theory the design problem is divided into two distinct parts—the approximation problem and the realization problem. In the following chapters these problems will be considered in detail. The approximation of the filter specifications by some aspect of a network function and the corresponding network function will be obtained. This function will be of the low-pass type or will be obtained from this type by frequency transformations. The realization problem, that of finding a network from the network function, will then be considered in a number of ways, using both passive and active networks.

EXERCISES

1.1. Given the following network functions, find the amplitude $|H(j\omega)|$ and loss $\alpha(\omega)$, and sketch both responses for $0 \le \omega \le 4$; determine which, if any, represent low-pass, high-pass, bandpass, band-reject, or all-pass filters:

(a) $H(s) = \dfrac{1}{s^2 + \sqrt{2}\,s + 1}$ *low*

(b) $H(s) = \dfrac{1}{s^3 + 2s^2 + 2s + 1}$

(c) $H(s) = \dfrac{s^2 - 2s + 2}{s^2 + 2s + 2}$

(d) $H(s) = \dfrac{s^3}{s^3 + 2s^2 + 2s + 1}$

(e) $H(s) = \dfrac{2s}{s^2 + 0.2s + 1}$

(f) $H(s) = \dfrac{s^2 + 1}{s^2 + 0.2s + 1}$ *high*

1.2. Find the time delay $T(\omega)$ and show that $T(0) = 1$, $T(1) = 12/13$ for the function

$$H(s) = \frac{3}{s^2 + 3s + 3}$$

Sketch $T(\omega)$ for $0 \le \omega \le 3$.

1.3. Find $T(\omega)$, $T(0)$, and $T(1)$ for the functions

(a) $H(s) = \dfrac{15}{s^3 + 6s^2 + 15s + 15}$ Ans. $T(1) = \dfrac{276}{277}$.

(b) $H(s) = \dfrac{105}{s^4 + 10s^3 + 45s^2 + 105s + 105}$ Ans. $T(1) = \dfrac{12{,}745}{12{,}746}$.

1.4. Show that

$$H(s) = \frac{2}{s^2 + 2s + 5}$$

is the transfer function of a low-pass filter, and find ω_c.

1.5. Show that the *RC* circuit shown in Fig. Ex. 1.5 is a low-pass filter and find ω_c, $\phi(\omega)$, and $T(\omega)$. (The network function is $V_2(s)/V_1(s)$.)

$$\text{Ans. } \omega_c = \frac{1}{RC}.$$

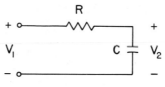

Figure Ex. 1.5.

2

Network Functions

2.1 NATURE OF THE NETWORK FUNCTIONS

In order to obtain, or *realize*, or *synthesize* a filter circuit that has a given amplitude or phase response, we shall first need to obtain a valid network function $H(s)$. In this chapter we shall consider the properties of such a function, how to find it from a given amplitude response, how to scale an impractical function to obtain one that is practical, and other topics related to the network functions.

The network function $H(s)$ may be the ratio of the transforms of two voltages, of two currents, of a voltage and a current, or of a current and a voltage. In any case, $H(s)$ may be theoretically obtained from the loop equations of the network. Suppose the network is of the type described in Sec. 1.2 and has n independent loops. Then we may write the loop equations in the form

$$Z_{11}I_1 + Z_{12}I_2 + \cdots + Z_{1n}I_n = V_1$$
$$Z_{21}I_1 + Z_{22}I_2 + \cdots + Z_{2n}I_n = V_2$$
$$\cdots \cdots \cdots \cdots \cdots \cdots \cdots \cdots \qquad (2.1)$$
$$Z_{n1}I_1 + Z_{n2}I_2 + \cdots + Z_{nn}I_n = V_n$$

where I_i is the ith loop current (in the s-domain) and V_j is the sum of the voltage sources, including terms due to initially stored energy, in the jth loop. The impedances Z_{ij} have the form

$$Z_{ij} = L_{ij}s + R_{ij} + \frac{C_{ij}^{-1}}{s} \tag{2.2}$$

where L_{ij}, R_{ij}, C_{ij}^{-1} are real numbers. Equation (2.1) may be written in matrix form as

$$Z_m I_m = V_m$$

where $Z_m = [Z_{ij}]$ is an $n \times n$ *loop impedance matrix*, $I_m = [I_1 I_2 \ldots I_n]^T$ is the loop current vector, and $V_m = [V_1 V_2 \ldots V_n]^T$ is the vector of voltage sources in the loops.

Suppose, for example, that the network function is given by

$$H(s) = Y_{pq} = \frac{I_p}{V_q} \tag{2.3}$$

implying that all the voltages V_i are zero except V_q. Then by Cramer's rule we obtain from (2.1)

$$I_p = \frac{\Delta_{qp} V_q}{\Delta}$$

where $\Delta = \det Z_m$ and Δ_{qp} is the cofactor of element Z_{qp} in Z_m. Thus we see, from the nature of Z_{ij}, that

$$H(s) = \frac{I_p}{V_q} = \frac{\Delta_{qp}}{\Delta} \tag{2.4}$$

is a *rational function* (a ratio of polynomials) whose coefficients are real and independent of the input V_q.

Voltage ratios may also be found using network determinants. For example, suppose it is desired to find

$$H(s) = \frac{V_p}{V_q}$$

where V_q is the only input and V_p is the voltage shown in Fig. 2.1. We have

$$H(s) = \frac{V_p}{V_q} = \frac{Z(I_r - I_s)}{V_q}$$
$$= Z(Y_{rq} - Y_{sq}) \tag{2.5}$$

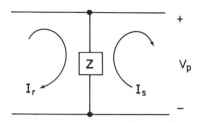

Figure 2.1. A network example.

which, as may be seen from (2.2) and (2.4), is also a rational function with real coefficients. A dual development may be carried out with the same results for current ratios and ratios of voltages to currents, using node equations.

The poles of a network function $H(s)$ are the values of s for which the function becomes infinite, and thus are either infinite (if the numerator exceeds the denominator in degree) or are the zeros of the denominator. In the latter case the poles are the natural frequencies of the network as may be seen from the inverse transform of $H(s)$. Therefore, in order to have stability, the poles cannot occur in the right half of the s-plane, nor can there be multiple poles on the $j\omega$-axis. The denominator polynomial in this case is said to be *Hurwitz*, a concept which we consider further in the next section.

The zeros of $H(s)$ are the values of s for which $H(s)$ is zero, and thus are either infinite (if the denominator exceeds the numerator in degree) or are the zeros of the numerator. They are also called *zeros of transmission* since evidently for these values of s the output $Y(s)$ is zero, and no transmission of energy has occurred from input to output.

2.2 THE HURWITZ TEST

As defined in Sec. 2.1, a Hurwitz polynomial

$$P(s) = M(s) + N(s) = a_n s^n + a_{n-1} s^{n-1} + \cdots + a_1 s + a_0 \qquad (2.6)$$

where M and N are the even- and odd-power terms respectively, is one with no right-half plane or multiple $j\omega$-axis zeros. If $P(s)$ has one or more simple $j\omega$-axis zeros in addition to possible left-half plane zeros then it is *modified* Hurwitz. If all its zeros are in the left-half plane (none on the $j\omega$-axis or in the right-half plane), then it is *strictly* Hurwitz. In this section we consider a test, called the Hurwitz test, to check a polynomial for the Hurwitz property without explicitly finding its zeros.

To begin with, if $P(s)$ is Hurwitz or modified Hurwitz, then its factors can have but three forms, namely

$$
\begin{aligned}
f_1 &= s + a, & a &\geq 0 \\
f_2 &= s^2 + b, & b &> 0 \\
f_3 &= s^2 + cs + d, & c, d &> 0
\end{aligned}
\tag{2.7}
$$

Thus $P(s)$ cannot have negative coefficients (unless all its coefficients are negative), and it cannot have missing powers between the highest- and lowest-degree terms, because there is no possibility of term cancellation when the product of such factors as (2.7) is formed. An exception to this latter statement occurs when $P(s)$ is either even or odd ($N = 0$ or $M = 0$) in which case all its factors except possibly a factor s are of the form of f_2. Such a polynomial is then modified Hurwitz. Finally, a_0 may be zero but a_0 and a_1 cannot both be zero, since this would yield a double zero at $s = 0$. Since the presence of a factor s is easily detected, as are negative or missing coefficients, we may assume that the polynomial to be tested is of the form (2.6) with $a_i > 0$; $i = 0$, $1, \ldots, n$.

The *Hurwitz* test for checking a polynomial $P = M + N$ of degree n for the Hurwitz property is as follows. (The interested reader may find a proof in [V].) Let $\psi = M/N$ or N/M so that ψ has a pole at infinity. Then if ψ may be expressed in the *continued* fraction form

$$
\psi = \alpha_1 s + \cfrac{1}{\alpha_2 s + \cfrac{1}{\ddots \cfrac{}{\cfrac{1}{\alpha_m s}}}}
\tag{2.8}
$$

where all $\alpha_i > 0$ and $m = n$, then $P(s)$ is *strictly* Hurwitz. If a negative α_i appears, then $P(s)$ is *not* Hurwitz. If all $\alpha_i > 0$ but $m < n$ (in which case *premature termination* has occurred), then $P(s)$ is the product of a Hurwitz polynomial of degree m and an even factor $W(s)$, common to both M and N. To further test $W(s)$, repeat the procedure with $\psi = W/W'$, where $W' = dW/ds$. (The factor $W(s)$ will emerge at the point of termination as common to the last divisor and dividend.) If $P(s)$ is even or odd, the testing begins with $\psi = P/P'$. In the case of W/W' or P/P', premature termination indicates that the polynomial .is not Hurwitz, for this implies a common even factor of multiplicity 2 or higher. Since W is even it can be at most modified Hurwitz, as discussed earlier. Factors like

$$W(s) = (s^2 - as + b)(s^2 + as + b)$$
$$= s^4 + (2b - a^2)s^2 + b^2$$

are of course not Hurwitz.

As an example, consider the obviously Hurwitz polynomial

$$P(s) = (s + 1)(s + 2)(s + 3)$$
$$= s^3 + 6s^2 + 11s + 6$$

We have

$$\psi = \frac{s^3 + 11s}{6s^2 + 6} = \frac{1}{6}s + \cfrac{1}{\cfrac{3}{5}s + \cfrac{1}{\cfrac{5}{3}s}} \tag{2.9}$$

Since the αs are all positive and premature termination does not occur, then $P(s)$ is strictly Hurwitz.

We may observe that the problem of obtaining the continued fraction is a process of divide, invert, divide, etc. Therefore a more compact method of obtaining (2.9) is as follows:

$$6s^2 + 6 \overline{)\, s^3 + 11s \,} \left(\tfrac{1}{6}s \right) \qquad \left(\alpha_1 = \tfrac{1}{6}\right)$$
$$\underline{s^3 + \ \ s}$$
$$10s \,\overline{)\, 6s^2 + 6 \,} \left(\tfrac{3}{5}s \right) \qquad \left(\alpha_2 = \tfrac{3}{5}\right)$$
$$\underline{6s^2}$$
$$6 \,\overline{)\, 10s \,} \left(\tfrac{5}{3}s \right) \qquad \left(\alpha_3 = \tfrac{5}{3}\right)$$
$$\underline{10s}$$

As another example, consider the obviously non-Hurwitz polynomial

$$P(s) = (s^2 + 1)^2(s + 2)$$
$$= s^5 + 2s^4 + 2s^3 + 4s^2 + s + 2$$

Performing the test we have

$$2s^4 + 4s^2 + 2 \,)\, s^5 + 2s^3 + s \,(\, s/2 \qquad (\alpha_1 = 1/2)$$
$$\underline{\underline{s^5 + 2s^3 + s}}$$

The process has terminated prematurely indicating that a $W(s)$ exists,
which must be common to

$$2s^4 + 4s^2 + 2 = 2(s^4 + 2s^2 + 1)$$

and

$$s^5 + 2s^3 + s = s(s^4 + 2s^2 + 1)$$

Thus $W(s) = s^4 + 2s^2 + 1 = (s^2 + 1)^2$ and therefore $P(s)$ is non-
Hurwitz (double zeros at $\pm j1$).

2.3 TWO-PORT NETWORKS AND PARAMETERS

A single-input, single-output electric filter is a *two-port* network (that is,
one with two pairs of terminals, an input, or primary, and an output,
or secondary, pair). The general form of a two-port network is shown
in Fig. 2.2.

The two currents I_1 and I_2 may be expressed in terms of V_1 and V_2
from the loop equations, which will be of the form of (2.1) with all

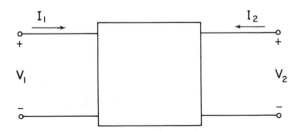

Figure 2.2. A two-port network.

$V_i = 0$ except V_1 and V_2. (We are considering the case of no internal independent sources.) Applying Cramer's rule we have

$$I_1 = \frac{\Delta_{11}}{\Delta}V_1 + \frac{\Delta_{21}}{\Delta}V_2$$

$$I_2 = \frac{\Delta_{12}}{\Delta}V_1 + \frac{\Delta_{22}}{\Delta}V_2 \tag{2.10}$$

where Δ is again det Z_m and Δ_{ij} is the cofactor of Z_{ji} in Z_m. Since the coefficients of V_i evidently have units of admittance we define

$$\frac{\Delta_{ij}}{\Delta} = y_{ji}; \qquad i, j = 1, 2 \tag{2.11}$$

so that (2.10) becomes

$$I_1 = y_{11}V_1 + y_{12}V_2$$

$$I_2 = y_{21}V_1 + y_{22}V_2 \tag{2.12}$$

Since by (2.12) we may write

$$y_{11} = \frac{I_1}{V_1}\Big|_{V_2=0}, \qquad y_{12} = \frac{I_1}{V_2}\Big|_{V_1=0}$$

$$y_{21} = \frac{I_2}{V_1}\Big|_{V_2=0}, \qquad y_{22} = \frac{I_2}{V_2}\Big|_{V_1=0} \tag{2.13}$$

we see that the y_{ij} may be obtained physically by making measurements under short-circuit conditions ($V_1 = 0$ or $V_2 = 0$). Therefore it is natural to call the y_{ij} the *short-circuit admittances* of the network.

By inverting (2.12) or by an analogous development using node equations we may write

$$V_1 = z_{11}I_1 + z_{12}I_2$$

$$V_2 = z_{21}I_1 + z_{22}I_2 \tag{2.14}$$

where z_{ij} are the *open-circuit impedances* given by

$$z_{11} = \frac{V_1}{I_1}\Big|_{I_2=0}, \qquad z_{12} = \frac{V_1}{I_2}\Big|_{I_1=0}$$

$$z_{21} = \frac{V_2}{I_1}\Big|_{I_2=0}, \qquad z_{22} = \frac{V_2}{I_2}\Big|_{I_1=0} \tag{2.15}$$

Evidently the z_{ij} and the y_{ij} are themselves network functions, being ratios of outputs to inputs under open-circuit or short-circuit conditions. A network function $Y(s)/X(s)$ is defined to be a *transfer function* if Y and X are measured at different ports, and a *driving-point* function if Y and X are measured at the same ports. Therefore $z_{11}, z_{22}, y_{11},$ and y_{22} are driving-point functions, whereas the others are transfer functions.

The two-port parameters z_{ij} and y_{ij} may be used to obtain general expressions for the various network functions $H(s)$. For example, if $H(s) = V_2/V_1$ for open-circuit secondary ($I_2 = 0$), we have from (2.14)

$$\frac{V_2}{V_1} = \frac{z_{21}}{z_{11}} \tag{2.16}$$

Similarly for short-circuit secondary ($V_2 = 0$), we have from (2.12)

$$\frac{I_2}{I_1} = \frac{y_{21}}{y_{11}} \tag{2.17}$$

Suppose now the secondary is terminated in a $1\ \Omega$ resistor, as shown in Fig. 2.3. Then $V_2 = -I_2$ and from the second of (2.12) we have

$$-V_2 = y_{21}V_1 + y_{22}V_2$$

from which we obtain

$$\frac{V_2}{V_1} = \frac{-y_{21}}{1 + y_{22}} = \frac{-I_2}{V_1} \tag{2.18}$$

In like manner, for Fig. 2.3, we may obtain from (2.14)

$$\frac{V_2}{I_1} = \frac{z_{21}}{1 + z_{22}} = \frac{-I_2}{I_1} \tag{2.19}$$

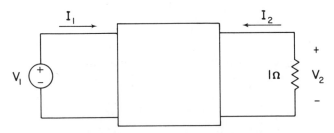

Figure 2.3. A two-port with resistive termination.

2.4 AN EXAMPLE

As an example utilizing the results of this chapter, let us consider the circuit of Fig. 2.4. We shall show that it is a low-pass filter and obtain its cutoff point and bandwidth. The transfer function is $H(s) = V_2(s)/V_1(s)$.

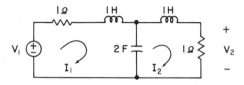

Figure 2.4. A low-pass filter.

Since there are no dependent sources present, the Z_{ij} of (2.2) may be written down by inspection. In this case Z_{ii} is the sum of the impedances in the ith loop ($i = 1, 2$), and Z_{ij}, $i \neq j$, is the negative of the impedances common to loops i and j. Therefore we have

$$\Delta = \begin{vmatrix} s + 1 + \dfrac{1}{2s} & -\dfrac{1}{2s} \\ -\dfrac{1}{2s} & s + 1 + \dfrac{1}{2s} \end{vmatrix}$$

$$= \left(s + 1 + \frac{1}{2s}\right)^2 - \left(\frac{1}{2s}\right)^2$$

$$= (s + 1)\left(s + 1 + \frac{1}{s}\right)$$

and

$$\Delta_{12} = \frac{1}{2s}$$

Therefore by (2.4), and the fact that $V_2 = I_2$, we have

$$\frac{V_2}{V_1} = \frac{I_2}{V_1} = \frac{1/2s}{(s + 1)(s + 1 + 1/s)}$$

or

$$H(s) = \frac{V_2}{V_1} = \frac{1/2}{s^3 + 2s^2 + 2s + 1} \qquad (2.20)$$

From this we obtain

$$H(j\omega) = \frac{1/2}{1 - 2\omega^2 + j(2\omega - \omega^3)}$$

or

$$|H(j\omega)| = \frac{1/2}{\sqrt{(1 - 2\omega^2)^2 + (2\omega - \omega^3)^2}}$$

$$= \frac{1/2}{\sqrt{1 + \omega^6}} \tag{2.21}$$

Evidently $|H(j\omega)|$ is a function which decreases monotonically as ω increases, and its maximum value of $\frac{1}{2}$ is attained at $\omega = 0$. Also we note that $\omega_c = 1$ since $|H(j1)| = (1/2)(1/\sqrt{2})$. Thus Fig. 2.4 is a low-pass filter with $\omega_c = 1$ and bandwidth $B = \omega_c = 1$. (As we shall see in Sec. 3.2, it is a low-pass *Butterworth filter* of *order* 3.) The amplitude response is shown in Fig. 2.5.

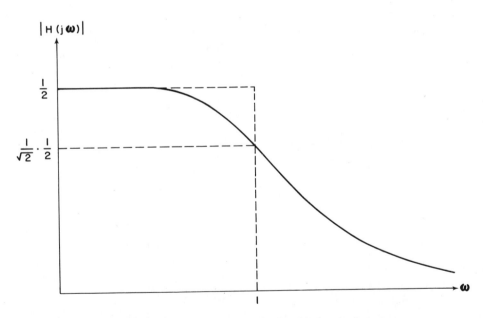

Figure 2.5. An amplitude response of a third-order low-pass Butterworth filter.

2.5 TRANSFER FUNCTIONS
FROM AMPLITUDE FUNCTIONS

As has been noted, in many cases the filter specifications are given in terms of the amplitude response. To obtain a network with a given amplitude response we shall first need to obtain a valid transfer function. Therefore, we shall consider in this section the general question of, given $|H(j\omega)|$, how do we find a suitable $H(s)$?

Since $H(s)$ is a ratio of polynomials with real coefficients, we know that

$$H^*(j\omega) = H(-j\omega)$$

where H^* is the complex conjugate of H. Also the given $|H(j\omega)|$ satisfies

$$|H(j\omega)|^2 = H(j\omega)H^*(j\omega)$$

and therefore

$$|H(j\omega)|^2 = H(j\omega)H(-j\omega)$$

This may also be written in the form

$$H(s)H(-s)\Big|_{s=j\omega} = |H(j\omega)|^2$$

or

$$H(s)H(-s) = |H(j\omega)|^2\Big|_{\substack{\omega^2=-s^2\\(j\omega=s)}} \tag{2.22}$$

Thus the problem is to separate out $H(s)$ from a given $H(s)H(-s)$. There is more than one way to do this, but the process is somewhat simplified by noting that if

$$H(s) = \frac{P(s)}{Q(s)}$$

then (2.22) becomes

$$\frac{P(s)}{Q(s)} \cdot \frac{P(-s)}{Q(-s)} = |H(j\omega)|^2\Big|_{\omega^2=-s^2}$$

Thus for stability $Q(s)$ must be the Hurwitz factor in $Q(s)Q(-s)$, and if no right-half plane zeros are desired, $P(s)$ must be the Hurwitz factor in $P(s)P(-s)$. Therefore we factor the numerator and denominator into, say $2n$ factors, retain the n Hurwitz factors and reject the n non-Hurwitz factors.

As an example, suppose we have

$$|H(j\omega)|^2 = \frac{4 + \omega^2}{1 + \omega^6}$$

(Note that an amplitude function is always a function of ω^2, since it is an even function.) Replacing ω^2 by $-s^2$ yields

$$H(s)H(-s) = \frac{4 - s^2}{1 - s^6}$$

$$= \frac{(2 - s)(2 + s)}{(1 - s^3)(1 + s^3)}$$

$$= \frac{(2 - s)(2 + s)}{(1 - s)(1 + s + s^2)(1 + s)(1 - s + s^2)}$$

If no right-half plane zeros are desired, then we have

$$H(s) = \frac{s + 2}{(s + 1)(s^2 + s + 1)} = \frac{s + 2}{s^2 + 2s^2 + 2s + 1}$$

the other factors constituting

$$H(-s) = \frac{-s + 2}{(-s + 1)(s^2 - s + 1)}$$

We should note at this point that driving-point functions cannot have right-half plane zeros, since their reciprocals are also driving-point functions. In the case of transfer functions, however, right-half plane zeros are admissible. Such a function is called a *nonminimum phase function* for reasons which we shall discuss in Sec. 7.2.

2.6 SCALING OF NETWORK FUNCTIONS

In Sec. 2.4 we considered a network which contained elements such as 1Ω, $1H$, and $2F$, which are delightful to a network analyst, but are distinctly less attractive to a network designer. Also the cutoff frequency of 1 rad/s or $1/2\pi$ Hz of the filter of Fig. 2.4 is not of any practical interest. We would have the best of both worlds if we could design on paper, circuits for 1 rad/s cutoff points using elements like 1Ω or $1F$, and use the results to construct the actual circuits for cutoff points like 1,000 Hz, using elements like 1 $k\Omega$ or 0.47 μF. Fortunately the concept of *network scaling* permits us to have it both ways.

Let us consider the transfer function $H(s) = Y_{pq} = I_p/V_q$ of Sec. 2.1, given in (2.4) by

$$Y'_{pq}(S) = \frac{I_p}{V_q} = \frac{\Delta'_{qp}(S)}{\Delta'(S)} \tag{2.23}$$

A typical entry in the determinants is given in (2.2) by

$$Z'_{ij}(S) = L'_{ij}S + R'_{ij} + \frac{1}{C'_{ij}S} \tag{2.24}$$

We are using S rather than s to represent the frequencies of the network functions before the network is scaled. Also, the primed quantities represent elements and determinants of the unscaled network.

We shall consider two types of network scaling, a *frequency scaling* in which S is replaced by s/k_f, and an *impedance scaling* in which Z'_{ij} becomes $Z_{ij} = k_i Z'_{ij}$. The factor k_f is the *frequency scale factor* and k_i is the *impedance scale factor*. Consider first frequency scaling, and let

$$Y_{pq}(s) = Y'_{pq}(s/k_f) = \frac{\Delta'_{qp}(s/k_f)}{\Delta_{qp}(s/k_f)} \tag{2.25}$$

Note that under this transformation, $s = k_f S$, so that if $s = j\omega$ corresponds to $S = j\Omega$, then $\omega = k_f \Omega$. Thus the values on the frequency axis have been multiplied by k_f, without affecting values on the vertical axis of a frequency response. Therefore, if the network had a cutoff point at 1 rad/s before scaling, then the scaled network has a cutoff of k_f rad/s.

The scaling of the network to effect the function scaling of (2.25) is quite simple. Replacing S by s/k_f in (2.24) yields

$$Z_{ij}(s) = Z'_{ij}(s/k_f) = \frac{L'_{ij}}{k_f}s + R'_{ij} + \frac{1}{\left(\frac{C'_{ij}}{k_f}\right)s}$$

$$= L_{ij}s + R_{ij} + \frac{1}{C_{ij}s}$$

where L_{ij}, R_{ij}, and C_{ij} represent the scaled network values. Therefore we have

$$L_{ij} = \frac{L'_{ij}}{k_f}$$

$$R_{ij} = R'_{ij} \qquad (2.26)$$

$$C_{ij} = \frac{C'_{ij}}{k_f}$$

and the scaling can be effected, in the case of passive elements, by leaving the resistances intact and dividing the capacitances and inductances by k_f.

If there are dependent sources present, the term Z'_{ij} is still of the form (2.24). For example, if the ith loop contains a dependent voltage source AV_k (A constant), controlled by the voltage V_k, then the ith loop equation is of the form

$$V_i = \sum_j Z'_{ij} I_j \pm AV_k$$
$$= \sum_j Z'_{ij} I_j \pm A \sum_j Z'_{kj} I_j$$
$$= \sum_j (Z'_{ij} \pm AZ'_{kj}) I_j$$
$$= \sum_j Z''_{ij} I_j$$

Thus since Z'_{ij} and Z'_{kj} are of the form (2.24), then Z''_{ij} is also of this form. If the dependent source AV_k is a current source, then we may find $n-1$ loop equations which do not contain AV_k, and the nth equation will be of the form

$$\sum_i \alpha_i I_i = AV_k$$

where $\alpha_i = 1$ or -1 if I_i flows through the dependent source (in one direction or the other) and $\alpha_i = 0$ otherwise. Thus $Z'_{ij} = \alpha_i/A$, a constant, which is a special case of (2.24). The reader may show in a similar way that for the other cases (a dependent voltage or current source AI_k, controlled by the current I_k), Z'_{ij} also has the form (2.24). In every case, frequency scaling leaves the gain constant A unchanged. (We shall consider only the case A constant.)

In the case of impedance scaling, we note that multiplying $Z'_{ij}(s)$ by

k_i is equivalent, in the passive case, to multiplying the inductances and resistances by k_i and dividing the capacitances by k_i. That is,

$$Z_{ij}(s) = k_i Z'_{ij}(s)$$

$$= k_i L'_{ij} s + k_i R'_{ij} + \frac{1}{\left(\dfrac{C'_{ij}}{k_i}\right)s}$$

$$= L_{ij} s + R_{ij} + \frac{1}{C_{ij} s}$$

In the case of dependent sources, dimensionless gain constants are unchanged, those with units of impedance are multiplied by k_i and those with units of admittance are divided by k_i. This may be seen from the dependent-source cases considered previously in the frequency scaling procedure.

Considering (2.23) we see that each element in both determinants has been multiplied by k_i. Therefore factoring out k_i from each row of each determinant yields

$$Y_{pq}(s) = \frac{k_i^{n-1} \Delta'_{qp}(s)}{k_i^n \Delta'(s)} = \frac{1}{k_i} Y'_{pq}(s)$$

Thus the function has been impedance scaled by a factor k_i, since Y'_{pq}, an admittance function has been scaled by a factor $1/k_i$.

The above process may also be shown to be valid in the case of the other network functions. In the case of dimensionless functions (ratios of voltages or ratios of currents), impedance scaling leaves them unchanged, but it multiplies impedance functions by k_i and divides admittance functions by k_i.

As an example, suppose we want to transform the filter of Fig. 2.4 to one with $\omega_c = 1{,}000$ rad/s, using a capacitor of 10 μF. Frequency scaling by a factor k_f and impedance scaling by a factor k_i requires that the 1Ω resistors be replaced by $k_i \Omega$ resistors, the 1H inductors be replaced by inductors of $L = k_i/k_f H$, and the 2F capacitor be replaced by $C = 2/k_i k_f F$. In addition, to change the cutoff from 1 rad/s to 1,000 rad/s requires that $k_f = 1{,}000$. Therefore we must have

$$C = \frac{2}{1{,}000\, k_i} = 10^{-5}$$

or $k_i = 200$. Thus we need resistors of 200Ω and inductors of $L = 200/1,000 = 0.2H$. The filter is shown in Fig. 2.6.

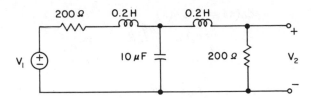

Figure 2.6. A low-pass filter with a cutoff frequency of 1,000 rad/s.

2.7 ANALYSIS AND SYNTHESIS

The problem of obtaining the network functions, considered in the previous sections, is a typical network *analysis* problem. That is, given the network N and the excitation $x(t)$ or its transform $X(s)$, find the response $y(t)$ or its transform $Y(s)$. An equivalent problem is that of finding the transfer function $H(s)$ since it may be used together with $X(s)$ to find $Y(s)$.

Another problem is that in which $x(t)$ and $y(t)$ are given, or equivalently $H(s)$ is given, and it is required to find the network N itself. This is the problem of network *synthesis* and is vastly more complicated than that of analysis. To begin with, in a linear network analysis problem, there are systematic ways to proceed, such as loop or node analysis, for example, and there is always a solution that is unique. For instance, in Fig. 2.4 loop analysis was used to obtain V_2/V_1, the answer is unique, and in the beginning there was no question but that an answer existed.

In the case of network synthesis the methods of attack are generally not as simple as say, writing a set of loop equations and proceeding systematically to solve them. The answer is never unique and indeed, there may not even be an answer. For example, suppose it is required to obtain passive one-port networks having input impedances given respectively by

$$Z_1 = \frac{s+1}{s^2+s+1} \tag{2.27}$$

and

$$Z_2 = \frac{s+2}{s^2+s+1} \tag{2.28}$$

These two functions are quite similar and have identical complexity. As we shall see in the next section it is impossible to synthesize Z_2 (that is, find a network) passively. On the other hand, Z_1 is readily synthesized, one solution being that of Fig. 2.7.

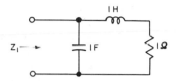

Figure 2.7. A realization of Eq. (2.27).

To see how Fig. 2.7 was arrived at and also to gain some appreciation of the many ways to approach a synthesis problem, we note that

$$Y_1 = \frac{1}{Z_1} = \frac{s^2 + s + 1}{s + 1}$$

which by long division is

$$Y_1 = s + \frac{1}{s + 1}$$

Since Y_1 is a sum of two admittances, s and $1/(s + 1)$, one approach is to rely on our analysis experience that the admittance of two parallel Ys is their sum. Thus a solution is a network having an admittance s (a capacitor of $1F$) in parallel with an admittance $1/(s + 1)$. This latter is an impedance of $s + 1$, which is realized by an impedance s (an inductor of $1H$) in series with an impedance of 1 (a resistor of 1Ω).

As another example, the impedance

$$Z = \frac{s + 1}{s(s + 2)} \tag{2.29}$$

may be written in partial fraction expansion form

$$Z = \frac{1}{2s} + \frac{1}{2(s + 2)}$$

and realized as two impedances in series, namely

$$Z_1 = \frac{1}{2s} \quad \text{and} \quad Z_2 = \frac{1}{2s + 4}$$

The first is realizable as a $2F$ capacitor and the second as an admittance of $2s + 4$, which is realizable as a $2F$ capacitor in parallel with a $\frac{1}{4}\Omega$ resistor. The result is shown in Fig. 2.8.

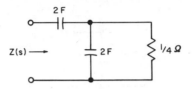

Figure 2.8. A realization of Eq. (2.29).

These simple examples illustrate how we may use our analysis experience to obtain, by a trial and error procedure, a network realization of a given $Z(s)$. Also from our experience we may suspect that no passive realization exists for the functions

$$Z_1 = s^2 + 1$$
$$Z_2 = s - 3 \qquad\qquad (2.30)$$
$$Z_3 = 2s + 3j$$

and save ourselves the trouble of searching for one. A more subtle example for which there is no passive realization is the function Z_2 of (2.28). Thus a natural question at this point is how can we be sure that a passive realization exists for a given $Z(s)$ before we embark on a long and possibly fruitless search for a network. We shall consider the answer to this question in the next section.

2.8 POSITIVE REAL FUNCTIONS

A function $Z(s)$ is said to be *positive real* if for *Re* $s > 0$,

(1) $Z(s)$ is real when s is real

(2) *Re* $Z(s) \geq 0$ $\qquad\qquad (2.31)$

These conditions were first obtained by Brune [Br] in 1931, when he showed that they were necessary and sufficient for $Z(s)$ to be the driving-point impedance of a passive network. For the interested reader, a derivation of the positive real conditions is presented in Appendix C.

As an example of a positive real function, which we abbreviate pr, let us consider the impedance

$$Z_1(s) = \frac{s+1}{s^2+s+1}$$

which was realized previously in Fig. 2.7. Replacing s by $\sigma + j\omega$ and rationalizing the denominator, we have $Re\, Z_1(s) = N/D$, where

$$D = |s^2+s+1|^2 > 0$$

and N is given by

$$N = (\sigma + 1)(\sigma^2 + \sigma + 1) + \omega^2\sigma$$

Evidently when $Re\, s = \sigma > 0$, we have $N \geq 0$ and hence $Re\, Z_1(s) \geq 0$. Also $Z_1(s)$ is real when s is real, and thus $Z_1(s)$ is pr.

In the case of

$$Z_2(s) = \frac{s+2}{s^2+s+1}$$

which was considered in (2.28), we have by the same procedure the same D as before and

$$N = (\sigma + 2)(\sigma^2 + \sigma + 1) + \omega^2(\sigma - 1)$$

which is negative when $\sigma = \frac{1}{2}$ and $\omega^2 > \frac{35}{4}$. Thus $Z_2(s)$ is not pr.

If $Z = r + jx$ is pr, then its reciprocal, $Y = 1/Z$, is pr. This is true since Y is evidently real when s is real (the reciprocal of a real number is real) and for $Re\, s > 0$,

$$Re\, Y = Re\, \frac{1}{r+jx} = \frac{r}{r^2+x^2} \geq 0$$

since in this case, $Re\, Z = r \geq 0$.

If Z is pr, then it can have no right-half-plane (RHP) pole, and any $j\omega$-axis pole must be simple with real positive residue. (The residue of a pole s_1 is the coefficient K in the term $K/(s - s_1)$ of the partial fraction expansion.) To show this, let us assume that Z has a RHP pole p of order m. That is,

$$Z = \frac{N(s)}{Q(s)(s-p)^m}$$

where $Q(p) \neq 0$. Expanding Z in partial fractions we have

$$Z = \frac{K_m}{(s-p)^m} + \frac{K_{m-1}}{(s-p)^{m-1}} + \cdots + \frac{K_1}{s-p} + Z_1 \qquad (2.32)$$

where Z_1 is the rest of the expansion. Suppose now that s is on a circle with center at $s = p$ and radius r sufficiently small so that the entire circle is in the right-half plane. That is,

$$s = p + re^{j\theta}$$

where $0 \leq \theta \leq 2\pi$. Suppose also that r is sufficiently small so that (2.32) may be approximated by

$$Z = \frac{K_m}{(s-p)^m} = \frac{K_m}{(re^{j\theta})^m}$$

If K_m is a complex number given by

$$K_m = Ke^{j\alpha}$$

where K is real and positive, then we have

$$Re\, Z = \frac{K}{r^m} \cos(\alpha - m\theta) \qquad (2.33)$$

As θ varies from 0 to 2π, $Re\, Z$ changes sign $2m$ times. Therefore for certain values of p in the RHP, $Re\, Z < 0$, which contradicts the assumption that Z is pr. Thus Z, if pr, can have no RHP pole.

We note however, that if p is on the $j\omega$-axis and s is in the RHP on a circle of radius r and center at p, then the range on θ is

$$\frac{-\pi}{2} \leq \theta \leq \frac{\pi}{2} \qquad (2.34)$$

Also by (2.33), $Re\, Z \geq 0$ if

$$\frac{-\pi}{2} \le \alpha - m\theta \le \frac{\pi}{2}$$

or

$$\frac{-\pi/2 + \alpha}{m} \le \theta \le \frac{\pi/2 + \alpha}{m}$$

This is consistent with (2.34) if $m = 1$ and $\alpha = 0$, which is the case of a simple $j\omega$-axis pole (order $m = 1$) with real positive residue. Thus a pr function may have simple $j\omega$-axis poles provided their residues are real and positive.

Clearly a pr function can have no RHP zeros either since its reciprocal is pr and can have no RHP poles. Also the numerator and denominator polynomials of a pr function cannot differ in degree by more than 1, for otherwise there would be a pole or zero at infinity (on the $j\omega$-axis) of order greater than 1. Similarly the lowest degree terms of the numerator and denominator cannot differ by more than 1 in degree, for this would be the case of a pole or zero at $s = 0$ of order greater than 1.

2.9 SUMMARY

The network functions are rational functions with real coefficients that are independent of the excitation. In the case of stable networks the poles of the functions may occur only in the left-half of the s-plane or on the $j\omega$-axis, in which case they must be simple. In the case of two-port networks, the network functions may be expressed generally in terms of the two-port parameters, such as the open-circuit impedances or the short-circuit admittances.

To fit a given set of specifications the network functions may be obtained directly or from some other function such as an amplitude function. As a rule, it is easier to work with a normalized transfer function because of its relative simplicity and the simple network element values which ensue. The concept of frequency and impedance scaling can then be used to make practical both the network realizations and the functions they perform.

In this chapter we have considered the nature of the network functions, which is an essential step in the synthesis procedure before studying the two basic problems of approximation and realization. In the next chapter we shall consider the problem of approximation, and obtain network functions for two well-known types of low-pass filters.

EXERCISES

2.1. Find $V_2(s)/V_1(s)$ for the network shown in Fig. Ex. 2.1.

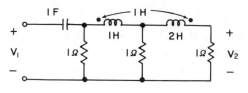

Figure Ex. 2.1.

2.2. Find $Z(s)$ for the network shown in Fig. Ex. 2.2.

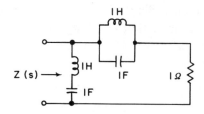

Figure Ex. 2.2.

2.3. Find the y-parameters for the network shown in Fig. Ex. 2.3.

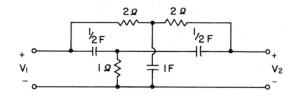

Figure Ex. 2.3.

2.4. For the network given in Fig. Ex. 2.4, show that the open-circuit impedances are

$$z_{11} = Z_a + Z_c$$
$$z_{12} = z_{21} = Z_c$$
$$z_{22} = Z_b + Z_c$$

2.5. For the network given in Fig. Ex. 2.5, show that the short-circuit admittances are

$$y_{11} = Y_a + Y_b$$
$$-y_{12} = -y_{21} = Y_b$$
$$y_{22} = Y_b + Y_c$$

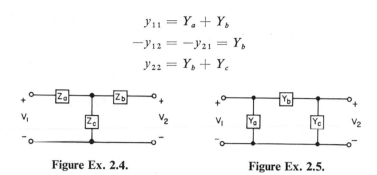

Figure Ex. 2.4.   Figure Ex. 2.5.

2.6. For the network given in Fig. Ex. 2.6, determine V_2/V_1 and show that the network is a low-pass filter with $\omega_c = 4$ rad/s. Scale the network so that it becomes a low-pass filter with $\omega_c = 40,000$ rad/s using a capacitor of 10 μF.

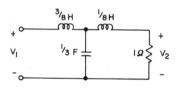

Figure Ex. 2.6.

2.7. For the network given in Fig. Ex. 2.7, determine V_2/V_1 and note its similarity to the function of Exercise 1.2. This is a *Bessel* filter of 2nd order to be considered later, and has a time delay that is nearly constant over $0 \le \omega \le 1$.

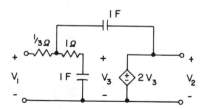

Figure Ex. 2.7.

2.8. For the network given in Fig. Ex. 2.8, show that

(a) $\dfrac{V_2}{V_s} = \dfrac{z_{21}R_2}{(R_1 + z_{11})(R_2 + z_{22}) - z_{21}z_{12}}$

(b) $Z(s) = \dfrac{z_{11}R_2 + \Delta_z}{z_{22} + R_2}$

where $z_{11}, z_{12}, z_{21}, z_{22}$ are the open-circuit impedances of the network N, and $\Delta_z = z_{11}z_{22} - z_{12}z_{21}$.

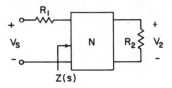

Figure Ex. 2.8.

2.9. The network given in Fig. Ex. 2.9 (a) is a *symmetrical* lattice. Show that its open-circuit parameters are given by

$$z_{11} = z_{22} = \tfrac{1}{2}(Z_b + Z_a)$$
$$z_{12} = z_{21} = \tfrac{1}{2}(Z_b - Z_a)$$

The network shown in (b) is a standard representation of the network of (a).

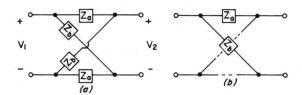

Figure Ex. 2.9.

2.10. Find V_2/I_1, V_1/I_1, and V_2/V_1 for the lattice shown in Fig. Ex. 2.10 (described in general in Exercise 2.9). Compare with $H(s)$ in Exercise 1.1 (c).

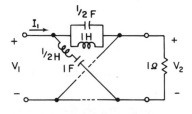

Figure Ex. 2.10.

2.11. Use the Hurwitz test to check the following polynomials:

(a) $s^3 + 7s^2 + 14s + 8$

(b) $s^3 + 2s^2 + 2s + 5$

(c) $s^4 + 3s^3 + 3s^2 + 3s + 2$

(d) $s^5 + s^4 + 2s^3 + 2s^2 + 5s + 5$

(e) $s^4 + s^2 + 1$

(f) $s^5 + 5s^3 + 4s$

2.12. Find V_2/V_1 for the circuit of Fig. 2.4 by assuming $V_2 = 1$, and finding in succession I_2, V_3 (the voltage across the $2F$ capacitor), I_1, and finally V_1. Since the circuit is linear, the quantities V_2 and V_1 are in the correct ratio.

2.13. Observe that the idea of Exercise 2.12 may be applied to the general linear ladder, shown in Fig. Ex. 2.13, and obtain the general iterative equations.

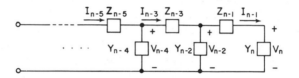

Figure Ex. 2.13.

Ans. $V_n = 1, \ I_{n-1} = Y_n$
$$V_{i-2} = Z_{i-1}I_{i-1} + V_i$$
$$I_{i-3} = Y_{i-2}V_{i-2} + I_{i-1}, \qquad i = n, n-1, \ldots$$

2.14. Given

$$|H(j\omega)|^2 = \frac{\omega^4 + 20\omega^2 + 64}{\omega^4 + 10\omega^2 + 9}$$

Find a realizable $H(s)$ with (a) minimum phase (b) nonminimum phase.

2.15. Find a realizable $H(s)$ if

(a) $|H(j\omega)|^2 = \dfrac{1 + \omega^2}{\omega^6 + 3\omega^4 - 3\omega^2 + 4}$

(b) $|H(j\omega)|^2 = \dfrac{(1 - \omega^2)^2}{\omega^6 + 3\omega^4 - 3\omega^2 + 4}$

2.16. (a) Show that if Z is pr, then kZ is pr, where k is a real, positive constant. (b) Show that if Z_i is pr, $i = 1, 2, \ldots, n$, then

$$Z = \sum_{i=1}^{n} Z_i$$

is pr.

2.17. Show that if $Z(s)$ is pr, then $Z(1/s)$ is pr.

2.18. Show that

$$Z = \frac{as}{s^2 + b}$$

is pr, where a and b are real, positive constants.

2.19. Synthesize

$$Z = \frac{s}{s^2 + 1} + \frac{2s}{s^2 + 4}$$

(Note that Z is pr by Exercises 2.16 and 2.18.)

2.20. The *maximum modulus theorem* from complex variable theory states that if $Z(s)$ is a rational function with no poles in or on the boundary of a closed region R, then the maximum value of $|Z(s)|$ in or on the boundary of R occurs on the boundary. Thus if $Z(s)$ has no RHP or $j\omega$-axis poles, then the maximum value of $|Z(s)|$ in the RHP or on the $j\omega$-axis occurs for some value of $s = j\omega$ (on the boundary of the RHP). Show that since $e^{-Z(s)}$ satisfies the conditions of the maximum modulus theorem if $Z(s)$ does, then $|e^{-Z(s)}| = e^{-Re\,Z(s)}$ and thus the minimum value of $Re\,Z(s)$ in the RHP or on the $j\omega$-axis occurs for some value of $s = j\omega$. Therefore if $Re\,Z(j\omega) \geq 0$, for $-\infty < \omega < \infty$, or since $Re\,Z(j\omega)$ is even, for $0 \leq \omega < \infty$, then $Re\,Z(s) \geq 0$ for all s in the RHP.

2.21. From the results of Exercise 2.20, show that $Z(s)$ is pr if
(a) $Z(s)$ is real when s is real and positive
(b) $Z(s)$ has no RHP poles
(c) Any $j\omega$-axis poles of $Z(s)$ are simple with real positive residues
(d) $Re\,Z(j\omega) \geq 0,\ 0 \leq \omega < \infty$
Suggestion: Write $Z = Z_1 + Z_2$, where Z_1 contains only the $j\omega$-axis poles (if any), and thus Z_2 by (b) satisfies the conditions of the maximum modulus theorem. By (c), Z_1 must contain only terms like $K_1 s$ or K_2/s which are pr, or terms like those of Exercise 2.18.

2.22. Using Exercise 2.21 show that Z_1 is pr and Z_2 is not pr in (2.27) and (2.28).

2.23. Synthesize

$$Z = \frac{s^3 + s^2 + 3s}{s^3 + 2s^2 + s + 2}$$

Suggestion: Factor the denominator and write $Z = Z_1 + Z_2$ as in the suggestion of Exercise 2.21.

2.24. The *hybrid* parameters h_{ij} and $g_{ij};\ i, j = 1, 2$, of a two-port network are defined by

$$V_1 = h_{11}I_1 + h_{12}V_2, \qquad I_2 = h_{21}I_1 + h_{22}V_2$$

and

$$I_1 = g_{11}V_1 + g_{12}I_2, \qquad V_2 = g_{21}V_1 + g_{22}I_2$$

Find both sets of hybrid parameters for the network of Exercise 2.3.

2.25. The *transmission* parameters, A, B, C, D, of a two-port network are defined by

$$V_1 = AV_2 - BI_2$$
$$I_1 = CV_2 - DI_2$$

Find A, B, C, D for the network of Exercise 2.3.

3

Approximation—
All Pole Filters

3.1 THE GENERAL LOW-PASS CASE

The ideal low-pass amplitude response, as we have seen in Chapter 1, has the form as shown in Fig. 3.1, where we have *normalized* the cutoff frequency to $\omega_c = 1$ rad/s. There is no loss in generality in the normalization since frequency scaling, discussed in Sec. 2.6, allows us to *denormalize* networks with $\omega_c = 1$ to obtain any desired cutoff point.

In this chapter and in Chapter 6 we shall consider ways to approximate the ideal response by a realizable transfer function. We shall show also that it is sufficient to study the low-pass filter case to obtain the general results in the high-pass, bandpass, and band-reject cases. This

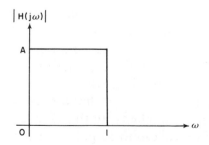

Figure 3.1. An ideal, normalized low-pass amplitude response.

is true because transformations may be performed directly on the low-pass, normalized networks to obtain the other types of filters. These transformations will be obtained in Chapter 4.

The amplitude function $|H(j\omega)|$ is an even function since it is the square root of the sum of the squares of the real and imaginary parts of $H(j\omega)$, which are respectively, even and odd functions. Therefore $|H(j\omega)|$ is a function of ω^2. Also, since $|H(j\omega)|$ is nonnegative it will be similar in form to $|H(j\omega)|^2$, which is easier to discuss since it does not involve a square root symbol.

A realizable $|H(j\omega)|^2$ which approximates the ideal case of Fig. 3.1 is given in general, for $A = 1$, by

$$|H(j\omega)|^2 = \frac{1}{1 + f(\omega^2)} \tag{3.1}$$

where

$$\begin{aligned} f(\omega^2) \gg 1, & \qquad \omega > 1 \\ 0 \leq f(\omega^2) \ll 1, & \qquad 0 \leq \omega < 1 \end{aligned} \tag{3.2}$$

This is evidently true since in the passband, $0 \leq \omega < 1$, we have $|H(j\omega)|^2 \approx 1$, and in the stopband, $\omega > 1$, we have $|H(j\omega)|^2 \approx 0$, when $f(\omega^2)$ is given by (3.2).

In the following sections we shall consider different functions $f(\omega^2)$ which satisfy (3.2), more or less. Generally, if $f(\omega^2)$ is a polynomial of degree $2n$ in ω (n in ω^2), then recalling the procedure in Sec. 2.5 of obtaining $H(s)$ from $|H(j\omega)|$, we see that $H(s)$ will be a constant divided by a polynomial of degree n in s. That is

$$H(s) = \frac{K}{Q(s)}$$

where $Q(s)$ is a polynomial of degree n. Such a function has all its zeros at infinity and all its poles at the finite zeros of $Q(s)$. Since $H(s)$ has no finite zeros, only finite poles, it is called an *all-pole* function, and the filter associated with it is sometimes referred to as an *all-pole* filter.

If $f(\omega^2)$ is a rational function, then $H(s)$ will also be a rational function. Thus $H(s)$ will have both finite poles and zeros as well as possible poles and zeros at infinity. In this chapter we shall concentrate on approximations which lead to all-pole filter functions, and consider in Chapter 6 the case of rational transfer functions.

3.2 BUTTERWORTH LOW-PASS FILTERS

One function suitable for use in (3.1) and (3.2) is

$$f(\omega^2) = \omega^{2n}; \qquad n = 1, 2, 3, \ldots \qquad (3.3)$$

first suggested by Butterworth [B]. In this case we have

$$|H(j\omega)| = \frac{1}{\sqrt{1 + \omega^{2n}}}; \qquad n = 1, 2, 3, \ldots \qquad (3.4)$$

which is defined as the amplitude response of the *nth-order Butterworth* filter. We note that the response is monotonically decreasing and thus it attains its maximum value, $|H(j\omega)|_{\max} = 1$, at $\omega = 0$. Also the cutoff point is the normalized value $\omega_c = 1$, since

$$|H(j1)| = \frac{1}{\sqrt{2}} = \frac{1}{\sqrt{2}}|H(j\omega)|_{\max}$$

The approximation (3.4) improves as n increases, since for $n_1 \gg n_2$, we have $\omega^{2n_1} \ll \omega^{2n_2}$ on $0 < \omega < 1$, and $\omega^{2n_1} \gg \omega^{2n_2}$ for $\omega > 1$. It is particularly good near $\omega = 0$, as we may see by expanding (3.4) using the binomial theorem; this results in

$$|H(j\omega)| = 1 - \frac{1}{2}\omega^{2n} + \frac{3}{8}\omega^{4n} - \frac{5}{16}\omega^{6n} + \frac{35}{128}\omega^{8n} - \cdots \qquad (3.5)$$

which is valid for ω near 0. The first $2n - 1$ derivatives of $|H(j\omega)|$ in (3.5) will contain a factor ω, and thus will be zero at $\omega = 0$. Therefore for n large, the function $|H(j\omega)|$ near $\omega = 0$ is exceedingly flat, or as it is defined in the literature, *maximally flat*. (See Exercise 3.7 for a formal definition of the term "maximally flat.")

For $\omega \gg 1$, the Butterworth amplitude function may be approximated by

$$|H(j\omega)| \approx \frac{1}{\omega^n}$$

with the loss in dB given by

$$\alpha_{dB}(\omega) \approx 20\log_{10}\omega^n = 20n\log_{10}\omega \qquad (3.6)$$

Thus if the loss is plotted versus ω in decades (a decade is the difference between two frequencies, one 10 times the other), then for large ω, the loss $\alpha_{dB}(\omega)$ has a slope of $20n$ dB/decade. The loss thus increases rapidly for large n, which indicates a good approximation to the ideal case.

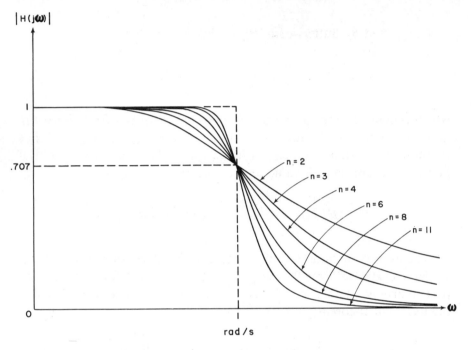

Figure 3.2. Butterworth amplitude responses.

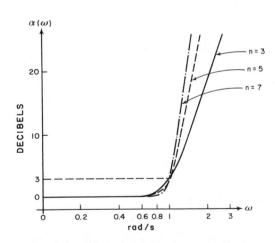

Figure 3.3. Butterworth loss curves.

Plots of $|H(j\omega)|$ and $\alpha_{dB}(\omega)$ are shown in Figs. 3.2 and 3.3 for various values of n. Evidently the approximation to the ideal amplitude improves as n increases. Finally, a response is shown in Fig. 3.4 of an actual sixth-order Butterworth circuit constructed in the laboratory.

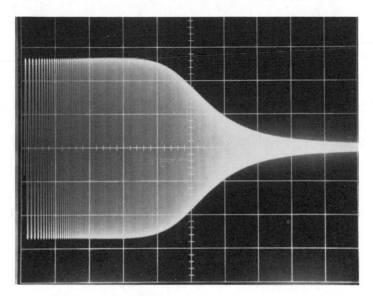

Figure 3.4. Response of an actual Butterworth filter of order 6.

3.3 BUTTERWORTH TRANSFER FUNCTIONS

Let us consider next the derivation of the Butterworth transfer function $H(s)$, whose amplitude is defined in (3.4). Replacing ω^2 by $-s^2$ in $|H(j\omega)|^2$ we have

$$H(s)H(-s) = \frac{1}{1 + (-s^2)^n}$$

so that

$$H(s) = \frac{1}{Q(s)} \qquad (3.7)$$

where $Q(s)$ is the Hurwitz polynomial satisfying

$$Q(s)Q(-s) = 1 + (-s^2)^n \qquad (3.8)$$

As an example, if $n = 2$, the case of the second-order Butterworth filter, (3.8) becomes

$$Q(s)Q(-s) = 1 + s^4$$

which we may write as

$$\begin{aligned} Q(s)Q(-s) &= s^4 + 2s^2 + 1 - 2s^2 \\ &= (s^2 + 1)^2 - (\sqrt{2}s)^2 \\ &= (s^2 + \sqrt{2}s + 1)(s^2 - \sqrt{2}s + 1) \end{aligned}$$

The first factor in the right member is the Hurwitz factor and is therefore $Q(s)$. Thus we have as the transfer function of the second-order Butterworth low-pass filter,

$$H(s) = \frac{K}{s^2 + \sqrt{2}s + 1} \qquad (3.9)$$

The constant K was normalized to 1 in (3.7), but in general may be any real number.

As another example, for $n = 3$, we have

$$Q(s)Q(-s) = 1 + \omega^6 \Big|_{\omega^2 = -s^2} = 1 - s^6$$

In this case $Q(s)$ has been obtained earlier in Sec. 2.5, where it was shown to be

$$Q(s) = s^3 + 2s^2 + 2s + 1$$

Therefore the transfer function of the third-order Butterworth low-pass filter is given by

$$H(s) = \frac{K}{s^3 + 2s^2 + 2s + 1} \qquad (3.10)$$

At this point we should note that Fig. 2.4, considered previously, is a third-order low-pass Butterworth filter. Also Exercises 1.1(a) and (b) dealt with second- and third-order Butterworth transfer functions.

In the general case of (3.8), we note that the zeros of $Q(s)$ are the left-half plane roots of the equation

$$1 + (-s^2)^n = 0$$

or

$$(-1)^n s^{2n} = -1 = e^{j(2k-1)\pi}; \qquad k = 0, 1, \ldots, 2n - 1 \qquad (3.11)$$

The kth root, $s_k = \sigma_k + j\omega_k$, satisfies

$$s_k^{2n} = e^{j(2k-1)\pi + j\pi n}$$

so that

$$s_k = \sigma_k + j\omega_k = e^{j(2k+n-1)\pi/2n} \qquad (3.12)$$

The real and imaginary parts of s_k are then

$$\sigma_k = \cos\frac{(2k + n - 1)\pi}{2n} = -\sin\frac{(2k - 1)\pi}{2n}$$

$$\omega_k = \sin\frac{(2k + n - 1)\pi}{2n} = \cos\frac{(2k - 1)\pi}{2n} \qquad (3.13)$$

Thus we see that since $\sigma_k^2 + \omega_k^2 = 1$, the zeros s_k are on a unit circle. This may also be seen from (3.12) (since the amplitude of s_k is 1), where it is also clear that the s_k are π/n radians apart. No s_k can occur on the $j\omega$-axis since this would require $\sigma_k = 0$ in (3.13) and hence $2k - 1$ to be an even integer. Therefore there are n left-half and n right-half plane zeros s_k. The n left-half plane values, which are the poles of $H(s)$, are given by (3.13) for $k = 1, 2, \ldots n$. This is true since for these values we have

$$0 < \frac{(2k - 1)\pi}{2n} < \pi$$

and thus σ_k is negative.

In the general case we may find the poles s_k from (3.13) and then find $Q(s)$ by forming the product

$$Q(s) = (s - s_1)(s - s_2) \cdots (s - s_n)$$
$$= s^n + b_{n-1}s^{n-1} + \cdots + b_1 s + b_0 \qquad (3.14)$$

These are the so-called *Butterworth polynomials* and are tabulated in Table A.1 of Appendix A for $n = 1, 2, \ldots 8$. In the even-order case, $Q(s)$ is shown as a product of second-degree factors in Table B.1 of Appendix *B* for $n = 2, 4, 6,$ and 8.

3.4 CHEBYSHEV LOW-PASS FILTERS

As we have seen in the previous section, the Butterworth low-pass amplitude response is very good in the vicinity of $\omega = 0$ and for large values of ω. However, as we may see from Fig. 3.2, the Butterworth characteristic is not very good in the vicinity of cutoff ($\omega = 1$). Evidently in the general case of (3.1), $|H(j\omega)|$ takes on its maximum value of 1 when $f(\omega^2) = 0$. In the case of the Butterworth filter we have lumped all the zeros of $f(\omega^2)$ at one point, namely $\omega = 0$, by choosing $f(\omega^2) = \omega^{2n}$. In this section we shall consider a filter, known as the *Chebyshev* filter, for which the zeros of $f(\omega^2)$ are spread out across the passband and consequently $|H(j\omega)|$ is forced to attain its maximum value of 1 at a number of passband points.

The Chebyshev low-pass filter has an amplitude defined, for the special case of (3.1), by

$$|H(j\omega)| = \frac{1}{\sqrt{1 + \epsilon^2 C_n^2(\omega)}}; \qquad n = 1, 2, 3, \ldots \qquad (3.15)$$

where ϵ is a constant and

$$C_n(\omega) = \cos\left(n \cos^{-1} \omega\right) \qquad (3.16)$$

is the *Chebyshev* polynomial of the first kind of degree n. (See, for example, [JJ].)

By (3.16) we see that

$$\begin{aligned} C_0(\omega) &= \cos 0 = 1 \\ C_1(\omega) &= \cos\left(\cos^{-1} \omega\right) = \omega \end{aligned} \qquad (3.17)$$

To obtain higher-degree polynomials, and to see that $C_n(\omega)$ is indeed a polynomial of degree n, let us make the substitution

$$\omega = \cos \theta \qquad (3.18)$$

so that

$$C_n(\omega) = \cos n\theta \qquad (3.19)$$

By means of trigonometric formulas we may write

$$\begin{aligned} C_{n+1}(\omega) &= \cos\left(n + 1\right)\theta = \cos n\theta \cos \theta - \sin n\theta \sin \theta \\ C_{n-1}(\omega) &= \cos\left(n - 1\right)\theta = \cos n\theta \cos \theta + \sin n\theta \sin \theta \end{aligned}$$

which when added yields

$$C_{n+1}(\omega) + C_{n-1}(\omega) = 2 \cos n\theta \cos \theta$$

or

$$C_{n+1}(\omega) = 2\omega C_n(\omega) - C_{n-1}(\omega) \tag{3.20}$$

Evidently if $C_{n-1}(\omega)$ and $C_n(\omega)$ are polynomials of degree $n - 1$ and n respectively, as they are for $n = 1$, then $C_{n+1}(\omega)$ is a polynomial of degree $n + 1$. Therefore by induction $C_n(\omega)$ in (3.16) is an nth degree polynomial.

We may obtain higher degree polynomials by means of (3.17) and (3.20). For example, $n = 1$ in (3.20) yields

$$C_2(\omega) = 2\omega C_1(\omega) - C_0(\omega)$$
$$= 2\omega^2 - 1$$

A list of Chebyshev polynomials for $n = 0, 1, \ldots 8$ is given in Table 3.1.

TABLE 3.1
Chebyshev Polynomials of the First Kind

n	$C_n(\omega)$
0	1
1	ω
2	$2\omega^2 - 1$
3	$4\omega^3 - 3\omega$
4	$8\omega^4 - 8\omega^2 + 1$
5	$16\omega^5 - 20\omega^3 + 5\omega$
6	$32\omega^6 - 48\omega^4 + 18\omega^2 - 1$
7	$64\omega^7 - 112\omega^5 + 56\omega^3 - 7\omega$
8	$128\omega^8 - 256\omega^6 + 160\omega^4 - 32\omega^2 + 1$

To see why the Chebyshev polynomials yield a good approximation in (3.15), let us consider some of their properties. Evidently we have

$$C_n(1) = \cos(n \cos^{-1} 1) = \cos 0 = 1 \tag{3.21}$$

Also, by reference to Fig. 3.5, we may write

$$C_n(-\omega) = \cos[n \cos^{-1}(-\omega)]$$
$$= \cos[n(\pi + \cos^{-1} \omega)]$$
$$= \cos n\pi \cos(n \cos^{-1} \omega) - \sin n\pi \sin(n \cos^{-1} \omega)$$

or

$$C_n(-\omega) = (-1)^n C_n(\omega) \tag{3.22}$$

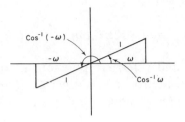

Figure 3.5. A sketch showing $\cos^{-1}(-\omega) = \pi + \cos^{-1}\omega$.

Thus $C_n(\omega)$ is even or odd according to whether n is even or odd, and therefore $C_n^2(\omega)$ is an even function.

The zeros of $C_n(\omega)$ are all real, distinct, and lie in the interval $-1 < \omega < 1$. This may be seen from (3.18) and (3.19), since $C_n(\omega) = 0$ requires $\theta = (2k-1)\pi/2n$, and thus the zeros ω_k are given by

$$\omega_k = \cos\frac{(2k-1)\pi}{2n}; \qquad k = 1, 2, \ldots, n$$

Also on $-1 \leq \omega \leq 1$ we have $|C_n(\omega)| \leq 1$ since $C_n(\omega)$ is the cosine of a real number in that case. Outside this range, on $|\omega| > 1$, $C_n(\omega)$ has no more zeros and thus is a monotonically increasing (or decreasing in the odd case of negative ω) function. A number of cases of $C_n(\omega)$ are plotted in Fig. 3.6.

Returning our attention to the amplitude response in (3.15) of the Chebyshev filter, we see that on $0 \leq \omega \leq 1$, $|H(j\omega)|$ attains its maxi-

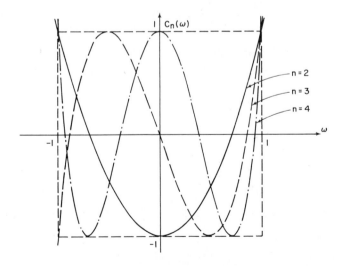

Figure 3.6. Chebyshev polynomials for $n = 2, 3, 4$.

mum value of 1 at the zeros of $C_n(\omega)$ and attains its minimum value of $1/\sqrt{1+\epsilon^2}$ at the points where $|C_n(\omega)|$ attains its maximum value of 1. Thus there are *ripples* in the passband, $0 \le \omega \le 1$, of *ripple width*

$$RW = 1 - \frac{1}{\sqrt{1+\epsilon^2}} \tag{3.23}$$

The Chebyshev filter is sometimes called an *equiripple* filter because the ripples in the passband are all equal in magnitude. This constant ripple width is often expressed in dB by calculating the loss in dB at the passband minima. This is given by

$$RW_{dB} = -20 \log_{10}\left(\frac{1}{\sqrt{1+\epsilon^2}}\right)$$

$$= 10 \log_{10}(1+\epsilon^2) \tag{3.24}$$

and used to characterize the Chebyshev filter. For example, a $\frac{1}{2}$ dB filter is one with ϵ such that $RW_{dB} = \frac{1}{2}$ (requiring $\epsilon = 0.3493$).

Outside the passband, $|\omega| > 1$, $|H(j\omega)|$ is monotonic decreasing, since as we have noted, $|C_n(\omega)|$ is monotonic increasing. The Chebyshev amplitude response therefore is a very suitable approximation to the ideal case.

As an example, the response of an actual $\frac{1}{2}$ dB Chebyshev filter of order 6, constructed in the laboratory, was shown previously in Fig. 1.3 of Chapter 1. This response is repeated in Fig. 3.7(b), where it may be compared with an actual response of a $\frac{1}{2}$ dB Chebyshev filter of order 2, shown in Fig. 3.7(a).

The actual response of Fig. 3.7(b) may be compared to the theoretical sixth-order Chebyshev response, shown for a given value of ϵ in Fig. 3.8. As is typical of even-order responses, $|H(j\omega)|$ has its minimum ripple value at $\omega = 0$. In the case of odd orders, the response is 1 at $\omega = 0$ because $C_n(0) = 0$. In every case, the amplitude is $1/\sqrt{1+\epsilon^2}$ at $\omega = 1$ because $C_n(1) = 1$.

For ω sufficiently large so that

$$\epsilon^2 |C_n(\omega)| \gg 1$$

we may approximate the amplitude by

$$|H(j\omega)| \approx \frac{1}{\epsilon C_n(\omega)}$$

Thus the loss is given by

$$\alpha(\omega) \approx 20 \log \epsilon + 20 \log C_n(\omega) \tag{3.25}$$

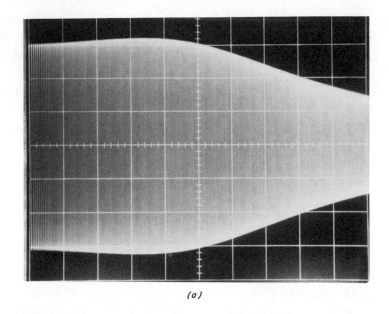

(a)

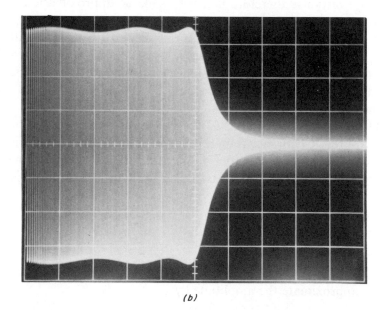

(b)

Figure 3.7. Actual $\frac{1}{2}$ dB Chebyshev low-pass responses of (a) order 2, and (b) order 6.

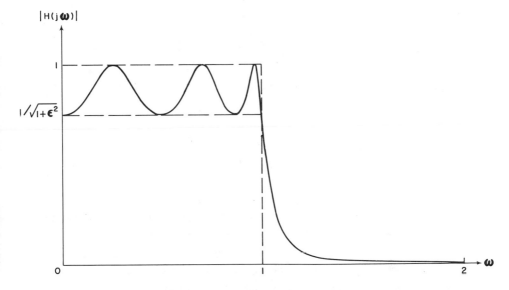

Figure 3.8. A sixth-order Chebyshev response.

The leading term in $C_n(\omega)$ is $2^{n-1}\omega^n$, as may be seen from repeated applications of (3.20), and for ω large, $C_n(\omega)$ may be approximated by the leading term. This further simplifies (3.25) to

$$\alpha(\omega) \approx 20 \log \epsilon + 6(n - 1) + 20n \log \omega \qquad (3.26)$$

Comparing this result with the Butterworth loss function of (3.6), we see that the Chebyshev loss function exceeds that of the Butterworth filter by $20 \log \epsilon + 6(n - 1)$. Since ϵ is usually a number between 0 and 1 to limit the ripple width, $\log \epsilon$ is negative. However, this is usually more than compensated for by the term $6(n - 1)$, except in the trivial case, $n = 1$.

We note that if $\epsilon = 1$, by (3.15) we have

$$|H(j1)| = \frac{1}{\sqrt{2}} = \frac{1}{\sqrt{2}}|H(j\omega)|_{\text{max}}$$

and thus $\omega_c = 1$. Also by (3.24), $\epsilon = 1$ corresponds to an approximate 3 dB *ripple channel*. (It is not exactly 3 dB since $\log_{10} 2$ is not exactly 0.3. For exactly 3 dB loss we need $\epsilon = 0.99763$.) For $0 < \epsilon < 1$, the cutoff point ω_c must be greater than 1 since the ripple channel terminates at $\omega = 1$. It may be shown (see Exercise 3.5) that the cutoff point is exactly

$$\omega_c = \cos h \left(\frac{1}{n} \cos h^{-1} \frac{1}{\epsilon} \right)$$

In Exercise 3.6 a list of cutoff points for various values of n and ripple widths is given.

3.5 EFFECT OF THE PARAMETERS
ON THE CHEBYSHEV RESPONSE

Evidently, increasing ϵ for a fixed n improves the stopband character-istics (while creating a larger ripple), since it increases the loss in (3.26). It may be seen too in (3.26) that for a fixed ϵ, increasing n increases the loss and improves the stopband characteristics also. In this case there are more ripples, of course. These two properties, increasing ϵ and increasing n to improve the stopband performance, are illustrated in Figs. 3.9 and 3.10. A comparison of two Chebyshev sixth-order re-sponses (0.1 dB and 0.969 dB) is shown with a sixth-order Butterworth response in Fig. 3.11. Clearly the cutoff and stopband features of the

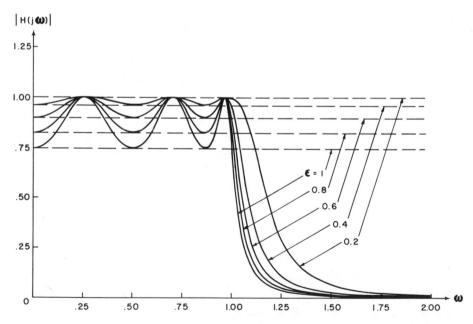

Figure 3.9. Variation of the Chebyshev response with ϵ, for $n = 6$.

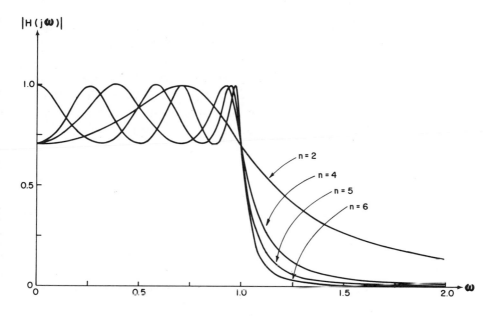

Figure 3.10. Variation of the Chebyshev response with n, for a 3 dB ripple.

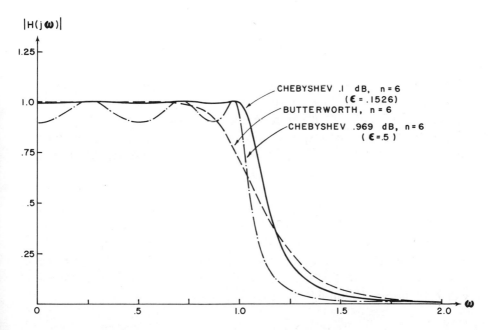

Figure 3.11. Comparison of the Butterworth and two Chebyshev responses for $n = 6$.

0.969 dB case are superior to those of the 0.1 dB case (though the opposite is true in the passband), and both are superior to the Butterworth response in this respect.

The Chebyshev filter is not only superior to the Butterworth at cutoff and in the stopband; it is, in fact, the optimum all-pole filter in this respect. Specifically, for a given allowable passband and stopband deviation, shown shaded in Fig. 3.12, of all the all-pole filters the Chebyshev has the shortest transition band, $1 < \omega < \omega_1$ [R].

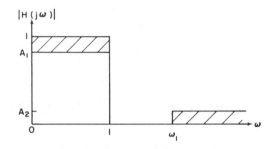

Figure 3.12. Passband and stopband tolerances.

3.6 COMPARISON OF BUTTERWORTH
AND CHEBYSHEV FILTERS

The Butterworth amplitude response, because of its maximally-flat property, is better near $\omega = 0$ than the Chebyshev amplitude response. However, in almost all other respects, as was noted in the previous section, the Chebyshev response is superior. (For excellent discussions of the merits of Butterworth and Chebyshev filters, the reader is referred to such standard references as [V], [W], and [Ba].)

To illustrate this last point, and also to give an example of a Chebyshev approximation problem, let us consider obtaining, for minimum n, both a Butterworth and a Chebyshev amplitude response to satisfy the requirements of Fig. 3.12 when $A_1 = 0.9$, $A_2 = 0.1$, and $\omega_1 = 2$. Since the cutoff point ω_c occurs when $|H(j\omega)| = 0.707$ in this case, we evidently must have $1 < \omega_c < 2$, and thus we must scale the Butterworth function, obtaining

$$|H(j\omega)|_B = \frac{1}{\sqrt{1 + (\omega/\omega_c)^{2n}}} \qquad (3.27)$$

At $\omega = 1$ we must have

$$\frac{1}{\sqrt{1 + (1/\omega_c)^{2n}}} \geq 0.9$$

or

$$(1/\omega_c)^{2n} \leq 19/81 \tag{3.28}$$

At $\omega = 2$ we must have

$$\frac{1}{\sqrt{1 + (2/\omega_c)^{2n}}} \leq 0.1$$

or

$$(1/\omega_c)^{2n} \geq \frac{99}{2^{2n}} \tag{3.29}$$

Therefore, from (3.28) and (3.29) we have

$$\frac{99}{2^{2n}} \leq (1/\omega_c)^{2n} \leq 19/81 \tag{3.30}$$

which is possible only if

$$\frac{99}{2^{2n}} \leq 19/81$$

or

$$2^n \geq \sqrt{81(99)/19} = 20.544 \tag{3.31}$$

Thus the minimum order is $n = 5$, for which the inequalities (3.30) may be written

$$1.15604 \leq \omega_c \leq 1.26318$$

If we choose $\omega_c = 1.2$, then the Butterworth response is given by

$$|H(j\omega)|_B = \frac{1}{\sqrt{1 + (\omega/1.2)^{10}}}$$

Thus we could construct a fifth-order normalized Butterworth filter and frequency scale it with a scale factor of 1.2.

Turning now to the Chebyshev filter, the specifications require a ripple width ≤ 0.1, or

$$\frac{1}{\sqrt{1 + \epsilon^2}} \geq 0.9 \tag{3.32}$$

and at $\omega = 2$, an amplitude

$$|H(j2)|_c = \frac{1}{\sqrt{1 + \epsilon^2 C_n^2(2)}} \leq 0.1 \qquad (3.33)$$

Since increasing ϵ or n improves the cutoff, for minimum n we should use the maximum allowable ϵ, which occurs when equality is chosen in (3.32). This results in

$$\epsilon^2 = 19/81 = 0.23457; \qquad \epsilon = 0.48432 \qquad (3.34)$$

With this value of ϵ, by (3.33) we must have

$$|C_n(2)| \geq \sqrt{81(99)/19} = 20.544 \qquad (3.35)$$

For $\omega = 2$, (3.20) becomes

$$C_{n+1}(2) = 4C_n(2) - C_{n-1}(2)$$

so that

$$\begin{aligned}
C_0(2) &= 1, \qquad C_1(2) = 2 \\
C_2(2) &= 4(2) - 1 = 7 \\
C_3(2) &= 4(7) - 2 = 26 \\
C_4(2) &= 4(26) - 7 = 97
\end{aligned} \qquad (3.36)$$

etc. Therefore (3.35) is satisfied for minimum order $n = 3$, and the Chebyshev response is given by

$$|H(j\omega)|_c = \frac{1}{\sqrt{1 + \frac{19}{81} C_3^2(\omega)}}$$

In this example, as is true in general, the Chebyshev amplitude is a superior fit to specifications such as those of Fig. 3.12. In the example, a third-order Chebyshev is as good as a fifth-order Butterworth, as far as Fig. 3.12 is concerned. In this case, as we shall see in Chapter 5, if both filters were constructed passively, the Butterworth would require 5 reactive elements (inductors and capacitors) whereas the Chebyshev would require only 3.

Finally, to further illustrate the Chebyshev filter, an amplitude response for $n = 15$ and $\epsilon = 0.1$ (0.0432 dB) is shown in Fig. 3.13.

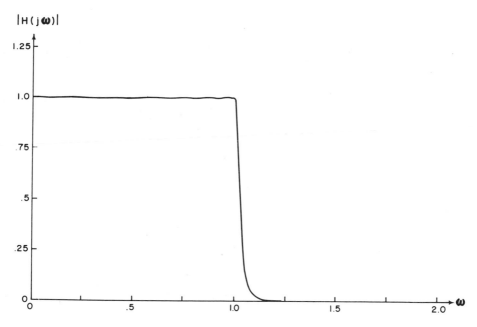

Figure 3.13. A fifteenth-order, 0.0432 dB ($\epsilon = 0.1$) Chebyshev response.

3.7 THE PHASE RESPONSES

Up to now we have focused our attention on the amplitude response $|H(j\omega)|$ of low-pass Butterworth and Chebyshev filters, and concluded that in most respects the Chebyshev is superior. As was pointed out earlier in Sec. 1.2, we are often also concerned with the phase response $\phi(\omega)$. As we shall see later, a phase response which is nearly linear over the passband is better than one which is very nonlinear.

As we make the amplitude response better, that is, more nearly ideal, unfortunately the accompanying phase response deteriorates. That is, the phase response becomes more nonlinear. We should then expect that the Butterworth phase response should be preferable to the Chebyshev phase response, and that lower-order Chebyshev phase responses are superior to those of higher order. This is precisely what happens and may be seen for a number of cases in Fig. 3.14. The sixth-order Butterworth response, shown by the broken line, is more nearly linear in the passband, $0 < \omega < 1$, than the Chebyshev responses of orders 3 through 7. Also the lower-order Chebyshev cases have better phase responses than the higher-order cases.

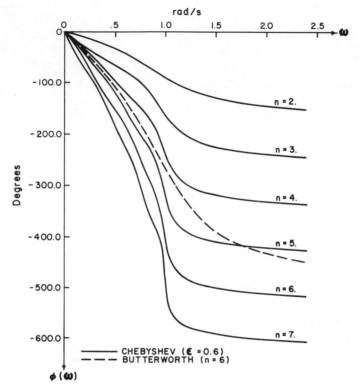

Figure 3.14. Butterworth and Chebyshev phase responses.

Thus if phase response is important as well as amplitude response, then the Chebyshev filter may not necessarily be better than the Butterworth. In case the phase response is the primary concern, as we shall see later, neither Butterworth nor Chebyshev filters should be used. A much better approximation then is the *Bessel* filter, to be considered later in Chapter 7

3.8 THE CHEBYSHEV FILTER TRANSFER FUNCTIONS

We note from (3.15) that the Chebyshev low-pass filter has an all-pole transfer function

$$H(s) = \frac{K}{Q(s)} \tag{3.37}$$

where by the results of Sec. 2.5 we have

$$Q(s)Q(-s) = 1 + \epsilon^2 C_n^2(\omega)$$

with $\omega^2 = -s^2$, or equivalently, $\omega = \pm js$. The poles of $H(s)$ are the left-half plane roots of the equation

$$1 + \epsilon^2 C_n^2(-js) = 0$$

or

$$C_n(-js) = \pm j/\epsilon \tag{3.38}$$

(We have chosen $\omega = -js$; since ω appears only in ω^2, it does not matter which sign we choose.)

Let us make the substitution

$$-js = \cos(u + jv) \tag{3.39}$$

which may be expanded in the form

$$-js = \cos u \cosh v - j \sin u \sinh v \tag{3.40}$$

since $\cos jv = \cosh v$ and $\sin jv = j \sinh v$. Also we have from (3.38) and (3.39),

$$C_n(-js) = \cos n(u + jv)$$
$$= \cos nu \cosh nv - j \sin nu \sinh nv = \pm j/\epsilon$$

Equating real and imaginary parts yields

$$\cos nu \cosh nv = 0$$
$$-\sin nu \sinh nv = \pm 1/\epsilon \tag{3.41}$$

Since for real v, $\cosh nv \neq 0$, we must have $\cos nu = 0$ in the first of (3.41), or

$$u = u_k = \frac{(2k - 1)\pi}{2n}; \qquad k = 1, 2, \ldots, 2n \tag{3.42}$$

Since $\sin u_k = \pm 1$, the second of (3.41) is satisfied if we take

$$\sinh nv = 1/\epsilon \tag{3.43}$$

or

$$v = \frac{1}{n} \sinh^{-1} 1/\epsilon \tag{3.44}$$

(We may vary k in (3.42) to satisfy the sign requirement in the second of (3.41).)

By (3.40) we may write

$$s_k = j \cos u \cosh v + \sin u \sinh v$$

where u and v are given by (3.42) and (3.44), and $k = 1, 2, \ldots, 2n$. The left-half plane members of this set, which are the poles of $H(s)$, are evidently given by $s_k = \sigma_k + j\omega_k$, where

$$\sigma_k = -\sin \frac{(2k-1)\pi}{2n} \sinh v$$

$$\omega_k = \cos \frac{(2k-1)\pi}{2n} \cosh v; \qquad k = 1, 2, \ldots, n$$

(3.45)

and v is given by (3.44).

From (3.45) we may show that the poles of the Chebyshev transfer function are located on an ellipse, given by

$$\frac{\sigma_k^2}{\sinh^2 v} + \frac{\omega_k^2}{\cosh^2 v} = 1 \qquad (3.46)$$

One may use this and the fact that the Butterworth poles are on a unit circle to develop a geometrical method of obtaining the Chebyshev poles from the Butterworth poles [V]. However, we shall obtain $\sinh v$ and $\cosh v$ in terms of n and ϵ and use (3.45) to obtain the Chebyshev transfer function.

By use of the identity

$$\cosh nv = \sqrt{1 + \sinh^2 nv}$$

and (3.43), we have

$$\cosh nv = \sqrt{1 + 1/\epsilon^2}$$

Also we have

$$e^{nv} = \cosh nv + \sinh nv$$
$$= \sqrt{1 + 1/\epsilon^2} + 1/\epsilon$$

from which

$$e^v = (\sqrt{1 + 1/\epsilon^2} + 1/\epsilon^2)^{1/n}$$

Finally, since

$$\sinh v = \frac{e^v - e^{-v}}{2}$$

$$\cosh v = \frac{e^v + e^{-v}}{2}$$

we have

$$\sinh v = \frac{1}{2}[(\sqrt{1 + 1/\epsilon^2} + 1/\epsilon)^{1/n} - (\sqrt{1 + 1/\epsilon^2} + 1/\epsilon)^{-1/n}]$$

$$\cosh v = \frac{1}{2}[(\sqrt{1 + 1/\epsilon^2} + 1/\epsilon)^{1/n} + (\sqrt{1 + 1/\epsilon^2} + 1/\epsilon)^{-1/n}]$$

(3.47)

The poles of $H(s)$ may now be found, for a given ϵ and n, by using (3.45) and (3.47).

As an example, let us find the transfer function of a Chebyshev filter with $n = 3$ and $\epsilon = 0.5$. We have

$$\sqrt{1 + 1/\epsilon^2} + 1/\epsilon = \sqrt{5} + 2 = 4.236068$$

so that (3.47) yields

$$\sinh v = \frac{1}{2}[(4.236068)^{1/3} - (4.236068)^{-1/3}] = 0.5$$

$$\cosh v = \frac{1}{2}[(4.236068)^{1/3} + (4.236068)^{-1/3}] = 1.118034$$

From (3.45) we have for $k = 1$,

$$\sigma_1 = -0.5 \sin \frac{\pi}{6} = -0.25$$

$$\omega_1 = 1.118034 \cos \frac{\pi}{6} = 0.968246$$

for $k = 2$,

$$\sigma_2 = -0.5 \sin \frac{\pi}{2} = -0.5$$

$$\omega_2 = 1.118034 \cos \frac{\pi}{2} = 0$$

and for $k = 3$,

$$\sigma_3 = -0.5 \sin \frac{5\pi}{6} = -0.25$$

$$\omega_3 = 1.118034 \cos \frac{5\pi}{6} = -0.968246$$

The denominator polynomial is then given by

$$Q(s) = (s + 0.5)(s + 0.25 - j0.968246)(s + 0.25 + j0.968246)$$
$$= (s + 0.5)(s^2 + 0.5s + 1)$$
$$= s^3 + s^2 + 1.25s + 0.5$$

and the transfer function is then

$$H(s) = \frac{K}{s^3 + s^2 + 1.25s + 0.5}$$

The denominator polynomials of the transfer functions of the normalized (ripple channel $0 \le \omega \le 1$) Chebyshev low-pass filter are tabulated in Appendix A for ripple widths of $0.1, 0.5, 1, 2$, and 3 dB and orders $n = 1, 2, \ldots, 8$. The quadratic factors of the denominator in the even-order cases $n = 2, 4, 6$, and 8 are compiled in Appendix B for these same ripple widths.

3.9 SUMMARY

To sum up the results of this chapter, the transfer function of an all-pole filter is of the form

$$H(s) = \frac{Gb_0}{s^n + b_{n-1}s^{n-1} + \cdots + b_1s + b_0} \tag{3.48}$$

where $b_0, b_1, \ldots, b_{n-1}$, and G are appropriately chosen constants. The Butterworth and Chebyshev low-pass filters have transfer functions of this type, differing only in the choice of the coefficients b_i. The coefficients for the normalized Butterworth and Chebyshev cases are given in Appendix A for $n = 1, 2, \ldots, 8$, and in Chebyshev cases, for ripple widths of $0.1, 0.5, 1, 2$, and 3 dB.

The *gain* of a low-pass filter is defined as $H(0)$, so that by (3.48) we see that the gain is G.

To obtain a low-pass filter of a certain type (Butterworth or Chebyshev) of order n, with gain G, and a specified cutoff point ω_c, we need only look up the b_i of (3.48) in Appendix A, and construct $H(s)$ for the specified G. (For Chebyshev ripple widths not given in the appendix, we may calculate the b_i, as described in the previous section.) Then we realize the network for the normalized cutoff point $\omega_c = 1$ rad/s, using techniques to be considered later, and frequency scale the normalized network to the desired cutoff point. It will also be necessary, in most cases, to impedance scale the network to obtain practical network element values.

EXERCISES

3.1. An *octave* is the difference between two frequencies, one twice the other. Show that for the Butterworth filter for large ω, the loss $\alpha(\omega)$ has a slope of approximately $6n$ dB/octave. (*Suggestion:* Note that $2^{\log_2 x} = x$.)

3.2. Use (3.13) to verify the Butterworth polynomial for $n = 4$ given in Appendix A.

3.3. An amplitude function of a low-pass filter has the following specifications:

$$|H(j\omega)|^2 \geq 0.5, \qquad 0 \leq \omega \leq 1$$
$$|H(j\omega)|^2 \leq 0.02, \qquad \omega \geq 2$$

Find $|H(j\omega)|^2$ for minimum n as (a) a Butterworth and (b) a Chebyshev filter function. Find $H(s)$ in each case.

$$\text{Ans. Chebyshev: } H(s) = \frac{K}{s^2 + \sqrt{\sqrt{2} - 1}s + 1/\sqrt{2}}$$

3.4. Show that the network given in Fig. Ex. 3.4 is a low-pass filter

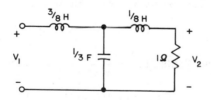

Figure Ex. 3.4.

$(H = V_2/V_1)$ and convert it to a low-pass filter with $\omega_c = 40,000$ rad/s using a capacitor of $10\mu F$.

$$\text{Ans. } \omega_c = 4 \text{ rad/s.}$$

3.5. Using the identity, $\omega = \cos \theta = \cosh j\theta$, and hence $j\theta = \cosh^{-1} \omega$, show that an alternate form of the Chebyshev polynomials is

$$C_n(\omega) = \cosh (n \cosh^{-1} \omega)$$

Use this result to show, by solving for ω_c in the equation

$$\epsilon C_n(\omega_c) = 1$$

that the cutoff point of the Chebyshev low-pass filter is given by

$$\omega_c = \cosh\left(\frac{1}{n}\cosh^{-1}\frac{1}{\epsilon}\right)$$

Note that if $0 < \epsilon < 1$, then $\omega_c > 1$.

3.6. Using the results of Exercise 3.5, verify some of the entries in the following table:

dB \ n	$f_{3\,\text{dB}}/f_c$ 2	3	4	5	6	7	8
0.1	1.943	1.389	1.213	1.135	1.093	1.068	1.052
0.5	1.390	1.168	1.093	1.059	1.041	1.030	1.023
1	1.218	1.095	1.053	1.034	1.023	1.017	1.013
2	1.074	1.033	1.018	1.012	1.008	1.006	1.005
3	1	1	1	1	1	1	1

Note that ϵ may be found for a given dB from (3.24) or from Appendix A.

3.7. Maximally flat functions are sometimes defined to be of the type

$$F(x) = \frac{a_0 + a_1x + a_2x^2 + \cdots + a_mx^m}{a_0 + a_1x + a_2x^2 + \cdots + a_mx^m + a_nx^n}$$

where $n > m$. (Note that $|H(j\omega)|^2$ of the low-pass Butterworth filter is of this form where $x = \omega^2$.) Show that

$$\left.\frac{d^k}{dx^k}F(x)\right|_{x=0} = 0; \qquad k = 1, 2, \ldots, n-1$$

Suggestion: Write $F(x) = \dfrac{p(x)}{p(x) + x^n}$ and note that

$$F'(0) = x^{n-1}R_1(x)\Big|_{x=0} = 0, \qquad n > 1,$$

$$F''(0) = x^{n-2}R_2(x)\Big|_{x=0} = 0, \qquad n > 2,$$

etc., where R_i is a rational function in x. Thus extend the argument to $F^{(k)}(0) = 0$, for $n > k$, or $k = 0, 1, \ldots, n-1$.

Alternate suggestion: By long division, obtain the Maclaurin series, valid near $x = 0$,

$$F(x) = \sum_{i=0}^{\infty} A_i x^i$$

and note that $A_i = 0$, $i = 1, 2, 3, \ldots, n-1$.

3.8. Find the minimum order of a Butterworth and of a Chebyshev filter which fits the specifications $A_1 = 0.95$, $A_2 = 0.05$, and $\omega_1 = 2$ of Fig. 3.12.

3.9. For a Chebyshev filter, find the largest ϵ and smallest n so that the ripple width is less than or equal to 0.1 (0.9152 dB) and $\omega_c \leq 1.25$ rad/s. (Here ω_c is the classical cutoff point or half-power point.)

3.10. A second-order Butterworth and a second-order Chebyshev filter satisfy the requirements $A_1 = 0.9$ and $A_2 = 0.05$ in Fig. 3.12, with ϵ equal to its maximum allowable value in the Chebyshev case. Find ω_1 for each filter.

4

Frequency
Transformations

4.1 INTRODUCTION

It is sufficient, for the synthesis of low-pass, high-pass, bandpass, and band-reject filters, to realize the normalized low-pass filter ($\omega_c = 1\,\text{rad}/s$). We shall establish this in this chapter by obtaining frequency transformations which may be applied directly to the normalized low-pass network to obtain these other types, for practical values of cutoff or center frequencies.

We shall start with the assumption that we have a normalized, low-pass network N_n, whose transfer function $H_n(S)$ has an amplitude response $|H_n(j\Omega)|$ which satisfactorily approximates the ideal response shown in Fig. 4.1. Then we shall seek a frequency transformation

$$S = F(s) \tag{4.1}$$

which applied to $H_n(S)$ results in the transfer function $H(s)$ of a *denormalized* network N having the required amplitude $|H(j\omega)|$. That is,

$$H(s) = H_n[F(s)] \tag{4.2}$$

and

$$|H(j\omega)| = |H_n(j\Omega)| \tag{4.3}$$

69

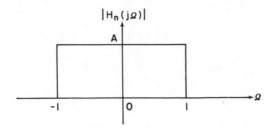

Figure 4.1. An ideal normalized low-pass response.

where

$$j\Omega = F(j\omega) \tag{4.4}$$

As we shall see, the transformation F may be applied directly to the normalized network N_n to obtain the network N, without the intermediate steps of (4.2) and (4.3).

We have already considered one example of a frequency transformation, that of frequency scaling considered in Sec. 2.6. In this case the transformation is

$$S = F(s) = \frac{s}{\omega_c} \tag{4.5}$$

which transforms the normalized low-pass network to a low-pass network with cutoff point ω_c. If ω_c is ω_0, a center frequency of a bandpass or band-reject filter, then (4.5) transforms a normalized such network (center frequency 1 rad/s) to one with center frequency ω_0. As we saw in Sec. 2.6, the normalized to denormalized transformation of (4.5) may be accomplished on the network N_n itself by replacing each inductance L and capacitance C by L/ω_c and C/ω_c respectively.

4.2 LOW-PASS TO HIGH-PASS TRANSFORMATION

Suppose the network N we are seeking is a high-pass filter with a prescribed cutoff ω_c. The ideal response of such a filter is shown in Fig. 4.2, where it may be seen that the transformation must map $\Omega = 0$ into $\omega = \infty$, and $\Omega = \pm 1$ into $\omega = \pm\omega_c$. Evidently the transformation required is the *low-pass* to *high-pass transformation* $F(s) = \omega_c/s$, or

$$S = \frac{\omega_c}{s} \tag{4.6}$$

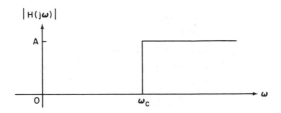

Figure 4.2. An ideal high-pass amplitude response.

since in this case we have, for $S = j\Omega$ corresponding to $s = j\omega$,

$$\Omega = -\frac{\omega_c}{\omega}$$

Thus $S = j0$ maps into $s = \pm j\infty$ and $S = \pm j1$ maps into $s = \mp j\omega_c$.

As discussed in Sec. 3.9, an all-pole normalized low-pass function is given in general by

$$H(s) = \frac{Gb_0}{s^n + b_{n-1}s^{n-1} + \cdots + b_1 s + b_0} \qquad (4.7)$$

where the b_i determine the type (Butterworth or Chebyshev, etc.) and G is the gain. Thus the high-pass transfer function obtained from the low-pass prototype function (4.7) by means of the transformation (4.6) is given, in the normalized case ($\omega_c = 1$), by

$$H(s) = \frac{Gb_0}{\dfrac{1}{s^n} + \dfrac{b_{n-1}}{s^{n-1}} + \cdots + \dfrac{b_1}{s} + b_0}$$

or

$$H(s) = \frac{Gs^n}{s^n + a_{n-1}s^{n-1} + \cdots + a_1 s + a_0} \qquad (4.8)$$

where, noting that $b_n = 1$, we have

$$a_{n-i} = b_i/b_0; \qquad i = 0, 1, 2, \ldots, n \qquad (4.9)$$

In the high-pass case the gain is defined as $\lim_{s\to\infty} H(s)$ and thus from (4.8) we see that the gain is G, the gain of the low-pass prototype filter.

Similar results may be derived using a low-pass prototype other than the all-pole type, such as the rational transfer function prototypes to be considered in Chapter 6.

The low-pass to high-pass transformation may be made directly on the low-pass network N_n to obtain the high-pass network N. The impedance of a resistor is unaffected by a frequency transformation, but in the case of an inductor L_n and a capacitor C_n, under the transformation (4.6) the impedances become respectively

$$Z_L(S) = L_n S = \frac{L_n \omega_c}{s} \triangleq \frac{1}{Cs}$$

$$Z_c(S) = \frac{1}{C_n S} = \frac{1}{C_n \omega_c / s} \triangleq Ls$$

Thus an inductance L_n in N_n becomes a capacitance $C = 1/L_n \omega_c$ in N, and a capacitance C_n in N_n becomes an inductance $L = 1/C_n \omega_c$ in N. A resistance R_n in N_n remains R_n in N. These transformations are summarized in Table 4.1 in Sec. 4.4.

An example of a high-pass amplitude is shown in Fig. 4.3, which is the response of an actual seventh-order $\frac{1}{2}$ dB Chebyshev filter.

As another example, the response of Fig. 1.4, shown earlier, is that of a sixth-order Butterworth high-pass filter.

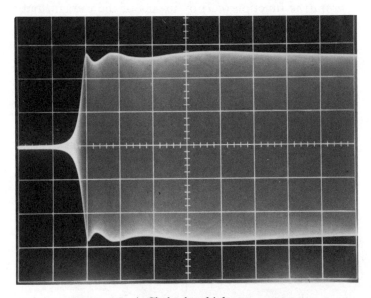

Figure 4.3. A Chebyshev high-pass response.

4.3 LOW-PASS TO BANDPASS TRANSFORMATION

To transform a low-pass prototype to a bandpass filter with center frequency ω_0 and bandwidth $B = \omega_U - \omega_L$, the transformation (4.1) must map the ideal low-pass response of Fig. 4.1 into the ideal bandpass response of Fig. 4.4. That is, (a) $S = j0$ must map into $s = j\omega_0$,

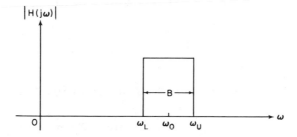

Figure 4.4. An ideal bandpass amplitude response.

(b) $S = j1$ must map into $s = j\omega_U$, and (c) $S = -j1$ must map into $s = j\omega_L$. Since $S = j\infty$ maps into $s = j\infty$, the function $F(s)$ must be an improper fraction,

$$S = F(s) = \frac{as^2 + bs + c}{ds + e} \tag{4.10}$$

(A quadratic over a linear factor is sufficient because of the number of constraints to be satisfied.)

Applying mapping (a) above to (4.10) results in

$$\frac{-a\omega_0^2 + jb\omega_0 + c}{jd\omega_0 + e} = 0$$

Equating the real and imaginary parts to 0 results in

$$c = a\omega_0^2, \qquad b = 0 \tag{4.11}$$

Applying mapping (b) and using (4.11) results in

$$j1(jd\omega_U + e) = -a\omega_U^2 + a\omega_0^2$$

or

$$e = 0, \qquad d = \frac{a(\omega_U^2 - \omega_0^2)}{\omega_U} \tag{4.12}$$

Mapping (c) results finally in

$$d = \frac{a(\omega_0^2 - \omega_L^2)}{\omega_L} \tag{4.13}$$

Equating the two values of d in (4.12) and (4.13) yields

$$\omega_L(\omega_U^2 - \omega_0^2) = \omega_U(\omega_0^2 - \omega_L^2)$$

or

$$\omega_0^2 = \frac{\omega_L\omega_U^2 + \omega_U\omega_L^2}{\omega_U + \omega_L} = \omega_L\omega_U \tag{4.14}$$

Thus under this transformation ω_0 is the geometric mean $\sqrt{\omega_L\omega_U}$ of ω_L and ω_U.

Using (4.11), (4.12), and (4.14), the transformation (4.10) becomes

$$S = \frac{a(s^2 + \omega_0^2)}{\left[\dfrac{a(\omega_U^2 - \omega_0^2)s}{\omega_U}\right]}$$

$$= \frac{s^2 + \omega_0^2}{(\omega_U - \omega_L)s}$$

or

$$S = \frac{s^2 + \omega_0^2}{Bs} \tag{4.15}$$

Thus a transfer function of a *2nth order* bandpass filter with center frequency ω_0 and bandwidth B is given by (4.7) with s replaced by S given in (4.15). The numerator will be a constant times s^n and the denominator will be a $2n$-th degree polynomial.

The *gain* of a bandpass filter is defined as the value of its transfer function $H(s)$ at the center frequency ω_0. Since $s = j\omega_0$, by (4.15), corresponds to $s = 0$ in (4.7), the gain of the bandpass filter is G, the gain of its low-pass prototype.

The *quality factor* Q is defined by

$$Q = \frac{\omega_0}{B} \tag{4.16}$$

so that in the normalized case, $\omega_0 = 1$ rad/s, we have, by (4.7) and (4.15),

$$H(s) = \frac{Gb_0}{s^n + b_{n-1}s^{n-1} + \cdots + b_1 s + b_0}\bigg|_{s \to [Q(s^2+1)/s]} \tag{4.17}$$

Thus a bandpass filter is a Butterworth or Chebyshev type depending on the choice of the coefficients b_i.

As an example, a second-order normalized bandpass filter with a specified Q and a gain G is described by

$$H(s) = \frac{G}{s+1}\bigg|_{s\to[Q(s^2+1)/s]}$$

or

$$H(s) = \frac{\dfrac{Gs}{Q}}{s^2 + \dfrac{1}{Q}s + 1} \tag{4.18}$$

(We have taken $b_0 = 1$ since $G/(s+1)$ is the transfer function of a normalized first-order low-pass filter with gain G.)

As examples of bandpass filter responses, we have plotted, for $Q = 1, 2, 5,$ and 10, fourth-order normalized Butterworth and 1 dB Chebyshev amplitude responses in Figs. 4.5 and 4.6 respectively. The

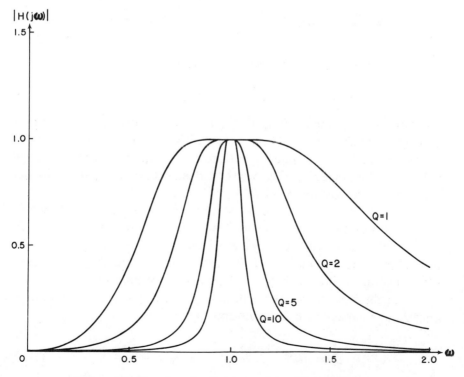

Figure 4.5. Fourth-order Butterworth bandpass responses.

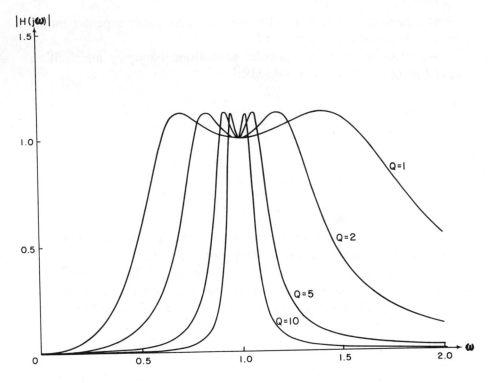

Figure 4.6. Fourth-order 1 dB Chebyshev bandpass responses.

functions were obtained from (4.17) for $n = 2$ with b_0, b_1, b_2, and b_3 appropriately chosen from Appendix A. The gain G is 1 in the Butterworth case and is such that the bottom of the passband ripple channel is 1 in the Chebyshev case. As we should expect from (4.16), the higher the Q the narrower the band B.

Examples of bandpass responses of actual circuits are shown in Figs. 4.7(a) and (b). The responses are those of fourth-order ($n = 2$) Butterworth and $\frac{1}{2}$ dB Chebyshev filters, both with $Q = 5$. (The horizontal scales are different.) The response of a fourth-order Butterworth bandpass filter with $Q = 10$ was shown earlier in Fig. 1.5.

As in the high-pass case we may scale the network N_n with inductance L_n and capacitance C_n by noting that

$$Z_L = L_n S = L_n \frac{s^2 + \omega_0^2}{Bs} = \frac{L_n}{B}s + \frac{\omega_0^2 L_n}{Bs}$$

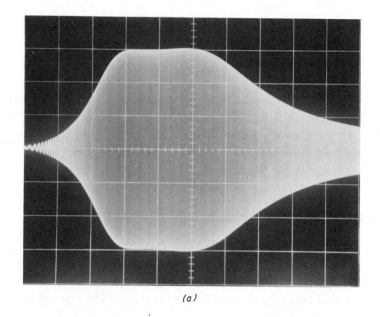

(a)

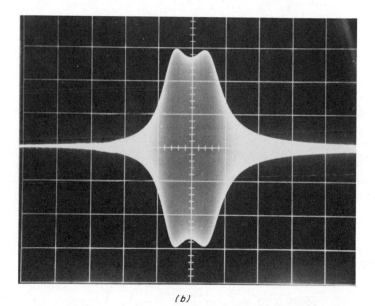

(b)

Figure 4.7. (a) Fourth-order Butterworth, and (b) fourth-order $\frac{1}{2}$ dB Chebyshev filter responses.

and

$$Z_c = \frac{1}{C_n S} = \frac{Bs}{C_n(s^2 + \omega_0^2)} = \frac{1}{\dfrac{C_n}{B}s + \dfrac{\omega_0^2 C_n}{Bs}}$$

Thus an inductance L_n in the low-pass prototype network transforms into an inductance $L = L_n/B$ in series with a capacitance $C = B/\omega_0^2 L_n$ in N, and a capacitance C_n transforms into a capacitance $C = C_n/B$ in parallel with an inductance $L = B/\omega_0^2 C_n$. Resistances remain unchanged. These transformations are tabulated in Table 4.1 in Sec. 4.4.

4.4 LOW-PASS TO BAND-REJECT TRANSFORMATION

In the case of a band-reject filter the transformation $F(s)$ must map Fig. 4.1 into Fig. 4.8. The result is a band-reject filter with center frequency ω_0 and width B of the band rejected. We may note that the transformation could be effected by first making a low-pass to high-pass

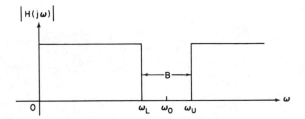

Figure 4.8. An ideal band-reject amplitude response.

transformation with $\omega_c = 1$, and then making the mappings (a), (b), and (c) of the low-pass to bandpass transformation. The resulting transformation then must be

$$S = \frac{Bs}{s^2 + \omega_0^2} \qquad (4.19)$$

We may directly transform the network N_n, as before, by noting that

$$Z_L = L_n S = L_n \frac{Bs}{s^2 + \omega_0^2} = \frac{1}{\dfrac{1}{BL_n}s + \dfrac{\omega_0^2}{BL_n s}}$$

$$Z_c = \frac{1}{C_n S} = \frac{s^2 + \omega_0^2}{BC_n s} = \frac{1}{BC_n}s + \frac{\omega_0^2}{BC_n s}$$

Thus an inductance L_n in N_n transforms into a capacitance $C = 1/BL_n$ in parallel with an inductance $L = BL_n/\omega_0^2$, and a capacitance C_n transforms into an inductance $L = 1/BC_n$ in series with a capacitance $C = BC_n/\omega_0^2$. These results, as well as those of the earlier sections, are tabulated in Table 4.1.

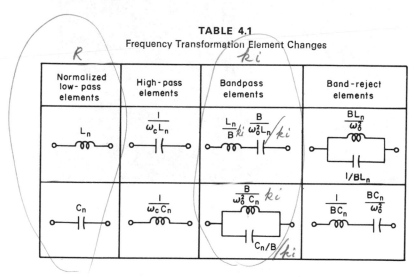

TABLE 4.1
Frequency Transformation Element Changes

The *gain* of a band-reject filter with transfer function $H(s)$ is defined as $H(0)$ or $\lim_{s\to\infty} H(s)$. By (4.19) and (4.7) we see that the gain is G, the gain of the low-pass prototype. Also, as in the bandpass case, we define $Q = \omega_0/B$, so that (4.19) may be written in the normalized case ($\omega_0 = 1$), as

$$S = \frac{s}{Q(s^2 + 1)} \qquad (4.20)$$

As an example, a second-order band-reject function is given by

$$H(s) = \frac{G(s^2 + 1)}{s^2 + \dfrac{1}{Q}s + 1} \qquad (4.21)$$

where the gain is G, the quality factor is Q, and the center frequency is $\omega_0 = 1$ rad/s. (Again we have normalized b_0 to 1.)

An example of the response of an actual band-reject filter is shown in Fig. 4.9. The response is that of a fourth-order 1 dB Chebyshev filter with $Q = 10$. This response was also shown earlier in Fig. 1.6.

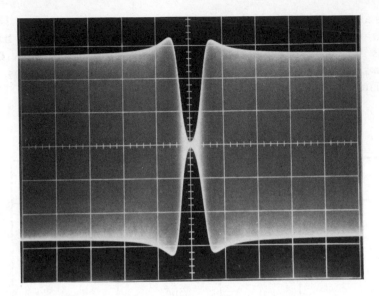

Figure 4.9. A fourth-order Chebyshev band-reject response.

4.5 AN EXAMPLE

To illustrate the passive synthesis of a filter obtained by a frequency transformation applied to a low-pass prototype, let us obtain a 6th-order Butterworth bandpass filter with a center frequency of $\omega_0 = 100{,}000$ rad/s and a Q of 10, from the low-pass third-order prototype of Fig. 2.4. This circuit, reproduced here as Fig. 4.10, has been shown earlier to be a normalized low-pass filter ($\omega_c = 1$ rad/s).

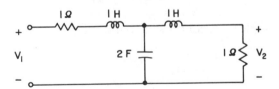

Figure 4.10. A third-order Butterworth low-pass prototype.

The low-pass to bandpass transformation of Table 4.1 indicates, since $B = \omega_0/Q = 10^4$, that the inductance $L_n = 1H$ is to be replaced by an inductance $L_n/B = 10^{-4}H$ in series with a capacitance $B/\omega_0^2 L_n = 10^{-6}F$, and the capacitance $C_n = 2F$ is to be replaced by an inductance

$B/\omega_0^2 C_n = 0.5 \times 10^{-6} H$ in parallel with a capacitance $C_n/B = 2 \times 10^{-4} F$. Making these replacements and impedance scaling the network of Fig. 4.10 by a factor k_i results in the network of Fig. 4.11. This is a bandpass filter with the required ω_0 and Q.

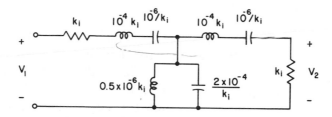

Figure 4.11. A Butterworth bandpass filter with variable impedance scaling.

The scale factor k_i should be chosen to make the elements as practical as possible. This will be difficult in the example we have chosen since the ratio of the two different sizes of inductances and of capacitances is 200 to 1. At the center frequency $\omega_0 = 10^5$ rad/s (15.9 kHz) typical values of inductance should range from $1H$ or so to tens of mH, and capacitances should be in the μF range. These are perhaps typical values in the frequency range of 1 Hz to 500 kHz, which is the range for which active filters, to be considered later, are superior to passive filters. From 500 kHz or so to 100 MHz and perhaps as high as 500 MHz, passive filters are superior, with typical inductances of mHs to 100s of μHs and capacitances in the μF to 100s of pF range. Probably above a frequency of 100 MHz, and certainly above 500 MHz, distributed networks should be used.

In the present example, let us compromise with a scale factor $k_i = 10^3$, resulting in the filter of Fig. 4.12. The range of element values

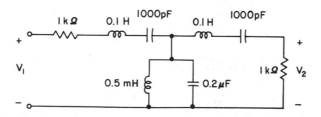

Figure 4.12. A sixth-order Butterworth bandpass filter.

is not optimum, to say the least, but is probably as good as we can do for the specified center frequency. Better element values may be obtained with an active realization, as we shall see.

4.6 RC:CR TRANSFORMATIONS

In the next chapter we shall consider methods of passively realizing filter functions. As we shall see, reactive elements (Ls and Cs) play a major role in passive synthesis. However, we shall be interested later in obtaining active realizations where the primary concern will be that of avoiding inductors. Thus the frequency transformations we have considered thus far will be of little help in transforming low-pass prototypes to high-pass filters, because the transformation $S = 1/s$ changes Cs to Ls. In this section we consider a transformation called an $RC:CR$ transformation [M-1], which effects the low-pass to high-pass transformation without introducing inductances.

Suppose the network N_n is an RC network with transfer function $V_p/V_q = ZI_p/V_q = ZY_{pq}$, where V_p is the voltage across element Z whose current is I_p. By (2.4) we may write the transfer function as

$$H(s) = V_p/V_q = \frac{Z\Delta_{qp}(s)}{\Delta(s)} \tag{4.22}$$

Typical entries in the determinants are, by (2.2),

$$Z_{ij}(s) = R_{ij} + \frac{C_{ij}^{-1}}{s} \tag{4.23}$$

since there are no inductors ($L_{ij} = 0$). Now perform the $RC:CR$ transformation, defined as replacing capacitances C_i by conductances $G_i = C_i$ and replacing conductances G_j by capacitances $C_j = G_j$. The new determinant entries are now

$$Z'_{ij}(s) = C_{ij}^{-1} + \frac{R_{ij}}{s} \tag{4.24}$$

with the new transfer function given by

$$H'(s) = \frac{Z'(s)\Delta'_{qp}(s)}{\Delta'(s)} \tag{4.25}$$

From (4.23) and (4.24) we see that

$$Z'_{ij}(s) = \frac{1}{s} Z_{ij}(1/s)$$

and since by the transformation we also have

$$Z'(s) = \frac{1}{s} Z(1/s)$$

then (4.25) becomes

$$H'(s) = \frac{1}{s} Z(1/s) \left[\frac{\frac{1}{s^{n-1}} \Delta_{qp}(1/s)}{\frac{1}{s^n} \Delta(1/s)} \right] \qquad (4.26)$$

(We are assuming the order of Δ to be *n*.) Simplifying the right member of (4.26) and comparing the result with (4.22) we see that

$$H'(s) = H(1/s) \qquad (4.27)$$

and the low-pass to high-pass transformation has been effected.

If the only elements in the network other than *R*s and *C*s are controlled sources with dimensionless gains, such as voltage-controlled voltage sources or current-controlled current sources, then the gains are unaffected by frequency transformations and (4.27) is still valid. (For the proof of this, see [M-2].) *Low pass → High pass*

As an example, let us consider the network of Exercise 2.7, which has a voltage-controlled voltage source with gain = 2. As the reader was asked to show, the transfer function is that of a second-order Bessel filter, given by

$$\frac{V_2}{V_1} = \frac{6}{s^2 + 3s + 3} \qquad (4.28)$$

Performing the *RC* : *CR* transformation on the circuit yields the network of Fig. 4.13. (The 1Ω resistors become $1F$ capacitors, and vice-versa,

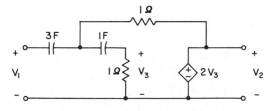

Figure 4.13. A high-pass circuit.

and the $\frac{1}{3}\Omega$ resistor becomes a $3F$ capacitor.) Analysis of Fig. 4.13 yields

$$\frac{V_2}{V_1} = \frac{6s^2}{3s^2 + 3s + 1}$$

which is a high-pass function equal to that of (4.28) with s replaced by $1/s$.

4.7 SUMMARY

By means of frequency transformations performed on the normalized low-pass transfer function we may obtain transfer functions for low-pass, high-pass, bandpass, and band-reject filters with practical values of cutoff or center frequencies. Alternately, we may perform the transformations on the normalized low-pass network itself to obtain the other filter networks. An RC : CR transformation may be used to transform certain low-pass circuits to high-pass circuits when it is important that the transformation not introduce inductors into the resulting network.

In the following chapters we may concentrate our attention on the synthesis of the low-pass normalized network, since we may obtain the others from it by frequency transformations. Chapters 5, 7, and 9 will be devoted to passive synthesis techniques and Chapters 10 and 11 to active synthesis procedures.

EXERCISES

4.1. Transform the circuit of Exercise 3.4 to a high-pass filter with $\omega_c = 10{,}000$ rad/s and capacitances of $1\mu F$ and $3\mu F$. (Note that the original circuit must have $\omega_c = 1$ for the frequency transformations to hold.)

4.2. Transform the circuit of Fig. 2.4 to a band-reject filter with $\omega_0 = 10{,}000$ rad/s and $Q = 10$. Impedance scale the resulting network so that the terminating resistances are $1k\Omega$.

4.3. Obtain the general transfer function

$$\frac{V_2}{V_1} = \frac{Gb_0 s^2/Q^2}{s^4 + \frac{b_1}{Q}s^3 + \left(2 + \frac{b_0}{Q^2}\right)s^2 + \frac{b_1}{Q}s + 1}$$

of a 4th-order bandpass filter with center frequency $\omega_0 = 1$ rad/s, gain G, and quality factor Q. The b_i are the low-pass prototype coefficients.

4.4. Obtain the general transfer function

$$\frac{V_2}{V_1} = \frac{G(s^2 + 1)^2}{s^4 + \frac{b_1}{b_0 Q}s^3 + \left(2 + \frac{1}{b_0 Q^2}\right)s^2 + \frac{b_1}{b_0 Q}s + 1}$$

of a 4th-order band-reject filter with given Q and G, for a center frequency $\omega_0 = 1$ rad/s. The b_i are the low-pass prototype coefficients.

4.5. Obtain a high-pass third-order Butterworth filter using the low-pass prototype of Fig. 2.4. Frequency and impedance scale the resulting network so that $\omega_c = 10,000$ rad/s and the terminating resistances are each $1 k\Omega$.

4.6. Show that the circuit given in Fig. Ex. 4.6 is a second-order normalized low-pass Butterworth filter with a gain of 1. Obtain a high-pass filter of the same type by means of the *RC: CR* transformation. (The transfer function is V_2/V_1).

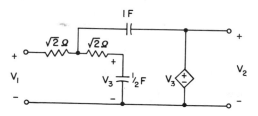

Figure Ex. 4.6.

5

Elements
of Passive Synthesis

5.1 THE TRANSFER FUNCTIONS

In this chapter we shall consider the realization or synthesis of filters from a given transfer function, using the passive elements, R, L, and C. We assume that the approximation problem has been solved and a transfer function has been obtained. The transfer functions we shall consider are those with strictly Hurwitz denominators and zeros on the $j\omega$-axis. We may note that this includes all the filter transfer functions we have considered thus far. The low-pass function of (4.7) has all its zeros at infinity, the high-pass function of (4.8) has all its zeros at the origin, the bandpass function of (4.17) has n zeros at the origin and n at infinity, and the band-reject function, defined by (4.7) with the substitution given by (4.20), has finite zeros on the $j\omega$-axis (all concentrated at the center, or notch frequency).

We shall consider in Chapter 6 more general transfer functions, but as we shall see, they too are of the type considered here. An exception is the all-pass filter, whose functions are considered in Chapter 7, along with some passive realizations.

The type of circuits we shall obtain are LC two-ports terminated in a single resistor at the output port. (*Doubly-terminated* circuits, in which there is also a source resistor, will be considered later in Chapter 9.)

Thus LC networks play a vital role in passive synthesis, and accordingly we shall devote the next two sections to a discussion of their properties.

5.2 *LC* DRIVING-POINT FUNCTIONS

Let us denote a driving-point LC impedance by $Z_{LC}(s)$. Since such a function is lossless we must have $Re\ Z_{LC}(j\omega) = 0$, and thus $Z_{LC}(s)$ must be a ratio of an odd polynomial $N(s)$ to an even polynomial $M(s)$, or vice versa. Since $N(s)$ is an even polynomial multiplied by the factor s, we may consider the properties of both M and N by considering the even polynomial $M(s)$.

We note that since both $Z_{LC}(s)$ and its reciprocal $Y_{LC}(s)$ are driving-point functions, both M and N must be Hurwitz polynomials. Thus the factors of M (and all those of N except the factor s) must be of the form $s^2 + a_1$, $s^2 + a_2, \ldots$, where $a_i > 0$ and $a_i \neq a_j$, $i \neq j$. For if $a_1 < 0$, then $s^2 = -a_1$ is positive, corresponding to a right- and a left-half plane zero, contradicting the Hurwitz property. This property is contradicted also if two or more of the a_i are equal (multiple $j\omega$-axis zeros). The only other possibility is a fourth-order factor such as

$$s^4 + as^2 + b = (s^2 + a_1)(s^2 + a_2)$$

where a_1 and a_2 are complex or pure imaginary. In this case the four zeros corresponding to the factor are complex. Suppose s_1 is one such zero. Since it is complex, it lies in the s-plane off the coordinate axes. Its negative, $-s_1$, is evidently also a zero, and since a and b are real, the conjugates s_1^* and $-s_1^*$ are also zeros. Thus we have a *quad* of zeros, two of which must be in the right-half plane, as shown in Fig. 5.1. This also contradicts the Hurwitz property.

Therefore we must have the general form

$$Z_{LC}(s) = \frac{K(s^2 + \omega_1^2)(s^2 + \omega_3^2) \cdots (s^2 + \omega_{2m-1}^2)}{s(s^2 + \omega_2^2)(s^2 + \omega_4^2) \cdots (s^2 + \omega_{2n}^2)} \qquad (5.1)$$

where $K > 0$, $\omega_1^2 \geq 0$, $\omega_i^2 > 0$, $i = 2, 3, \ldots$, and $\omega_i \neq \omega_j$, $i \neq j$. We are allowing the possibility $\omega_1^2 = 0$ so that we may include both possibilities $Z_{LC} = M/N$ (when $\omega_1^2 > 0$) and $Z_{LC} = N/M$ (when $\omega_1^2 = 0$). We also observe that since Z_{LC} is a pr function, the degrees of M and N cannot differ by more than 1. Since their degrees cannot be equal, they

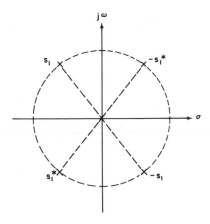

Figure 5.1. A quad of zeros corresponding to a certain quartic factor.

must then differ by exactly 1. This means that either we have $m = n$ or $m = n + 1$.

Other properties of $Z_{LC}(s)$ may be found from a partial fraction expansion. Since by (5.1) we see that Z_{LC} has a possible pole at $s = 0$ (in case $\omega_i^2 > 0$) and at infinity (if $m = n + 1$), the expansion must be of the form

$$Z_{LC}(s) = A_\infty s + \frac{A_0}{s} + \sum_{k=1}^{n} \left[\frac{a_k}{s - j\omega_{2k}} + \frac{a_k^*}{s + j\omega_{2k}} \right] \qquad (5.2)$$

Since Z_{LC} is pr we know that the residues must be real and nonnegative, so that $a_k = a_k^* > 0$ and therefore

$$Z_{LC}(s) = A_\infty s + \frac{A_0}{s} + \sum_{k=1}^{n} \frac{A_k s}{s^2 + \omega_{2k}^2} \qquad (5.3)$$

where $A_k = 2a_k$. Thus A_∞, A_0, and A_k, $k = 1, 2, \ldots, n$, are nonnegative real numbers.

If $s = j\omega$ in (5.3) we have

$$Z_{LC}(j\omega) = jX(\omega) = j\left[A_\infty \omega - \frac{A_0}{\omega} + \sum_{k=1}^{n} \frac{A_k \omega}{\omega_{2k}^2 - \omega^2} \right] \qquad (5.4)$$

from which we obtain

$$\frac{dX(\omega)}{d\omega} = A_\infty + \frac{A_0}{\omega^2} + \sum_{k=1}^{n} \frac{A_k(\omega^2 + \omega_{2k}^2)}{(\omega_{2k}^2 - \omega^2)^2} \qquad (5.5)$$

Since all the terms in the right member are nonnegative we conclude that

$$\frac{dX(\omega)}{d\omega} > 0, \qquad -\infty < \omega < \infty \qquad (5.6)$$

Therefore since $X(\omega)$ is continuous except at its poles and $dX/d\omega$ is everywhere positive, it follows that between every two poles there is a zero and between every two zeros there is a pole. (There is no way for a continuous function, as $X(\omega)$ is between two poles, with monotonic slope to pass through zero more than once.) A plot of one possible $X(\omega)$ is shown in Fig. 5.2. Other possibilities include a zero at $s = 0$

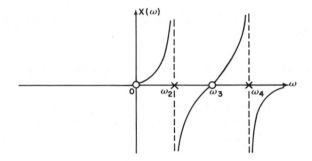

Figure 5.2. A plot of $X(\omega)$ versus ω.

and a pole at infinity, a pole at zero and a zero at infinity, and a pole at zero and a pole at infinity. This property of having the poles and zeros alternate or *interlace* on the $j\omega$-axis leads us to the condition,

$$0 \leq \omega_1 < \omega_2 < \omega_3 < \cdots \qquad (5.7)$$

The plot of $X(\omega)$ in Fig. 5.2 is not shown for $\omega < 0$, but could be added easily since $X(\omega)$ is an odd function.

5.3 REALIZATIONS OF *LC* FUNCTIONS

From our knowledge of network analysis we could realize Z_{LC} in the partial fraction form of (5.3) by interpreting the terms as impedances of elements connected in series. The first two terms represent respectively an inductance $L = A_\infty$ and a capacitance $C = 1/A_0$. A typical term in the summation may be written

$$\frac{A_k s}{s^2 + \omega_{2k}^2} = \frac{1}{\dfrac{s}{A_k} + \dfrac{\omega_{2k}^2}{A_k s}}$$

and thus interpreted as an inductance $L_k = A_k/\omega_{2k}^2$ in parallel with a capacitance $C_k = 1/A_k$. The network thus obtained is shown in Fig. 5.3 and is called the *Foster* 1 network in honor of Foster [F] who first obtained it in 1924.

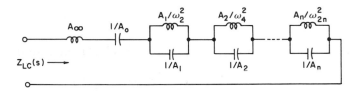

Figure 5.3. The Foster 1 *LC* network.

We may describe $Z_{LC}(s)$ as the ratio of two polynomials, one even and one odd, with poles and zeros which interlace on the $j\omega$-axis and a positive gain constant K, as shown in (5.1). Since this description also fits $Y_{LC}(s)$, the reciprocal of $Z_{LC}(s)$, we conclude that both functions have identical forms. Thus Y_{LC} also has the form given by (5.3), but its interpretation must be that terms added represent parallel admittances. The network obtained in this manner, using Y_{LC} instead of Z_{LC}, is due to Cauer* and is known as the *Foster* 2 network. It is further considered in Exercise 5.3.

Let us now consider two other general realizations of $Z_{LC}(s)$, known as the *Cauer* 1 and *Cauer* 2 networks, both given originally by Cauer. The Cauer 1 network is obtained from a continued fraction expansion "about infinity," as follows. If in (5.3), $A_\infty \neq 0$, then we may write

$$Z_{LC}(s) = \alpha_1 s + Z_1(s) \qquad (5.8)$$

where $\alpha_1 = A_\infty$ and

$$Z_1(s) = \frac{A_0}{s} + \sum_{k=1}^{n} \frac{A_k s}{s^2 + \omega_{2k}^2} \qquad (5.9)$$

*Wilhelm A. E. Cauer, 1900–1945, was one of the great pioneers of network theory.

(If $A_\infty = 0$, then $Y_{LC}(s)$ may be written in this form with Y_1 instead of Z_1.) Now Z_1 is an LC impedance without a pole at infinity. (The operation of (5.8) is called "removing the pole at infinity.") Since Z_1 has no pole at infinity, its reciprocal Y_1 must have, so that Y_1 has the form

$$Y_1(s) = \alpha_2 s + Y_2 \qquad (5.10)$$

where Y_2 is an LC admittance with the form of Z_1 in (5.9). Continuing the argument, we have

$$\frac{1}{Y_2} = Z_2 = \alpha_3 s + Z_3 \qquad (5.11)$$

and so forth. Summing up Eqs. (5.8) through (5.11), etc., we may write

$$Z_{LC}(s) = \alpha_1 s + \cfrac{1}{\alpha_2 s + \cfrac{1}{\alpha_3 s + \cfrac{1}{\ddots \cfrac{}{\cfrac{1}{\alpha_N s}}}}} \qquad (5.12)$$

We may observe here that $Z_{LC} = [M(s)/N(s)]^{\pm 1}$ and is pr. Also since 1 is pr and the sum of two pr functions is pr, then $1 + M/N = (M + N)/N$ is pr. Therefore $M + N$ is the numerator of a pr function and is thus a Hurwitz polynomial. Thus we could have obtained the continued fraction expansion (5.12) by performing the Hurwitz test on the function $(M/N)^{\pm 1}$. It is tacitly assumed that there are no common factors in M and N, so that the process will not terminate prematurely.

The Cauer 1 network may be drawn by noting that Z_{LC} in (5.12) is an inductance α_1 in series with the remainder, which is a capacitance α_2 in parallel with the rest, etc. That is, the Cauer 1 network is a ladder network with inductances $\alpha_1, \alpha_3, \ldots,$ in the series arms and capacitances $\alpha_2, \alpha_4, \ldots,$ in the shunt arms, as shown in Fig. 5.4. If $Z_{LC}(s)$ has no pole at infinity, then $Y_{LC}(s)$ is the function expanded as in (5.12), so that the first element is a shunt capacitor with capacitance α_1, etc. In either case the ladder network will end with a series inductor or a shunt capacitor, depending on the number of elements N.

The number N of poles of $Z_{LC}(s) = [M(s)/N(s)]^{\pm 1}$ is equal to the degree of $M(s)$ or $N(s)$, whichever is higher. It is evident from (5.3) that

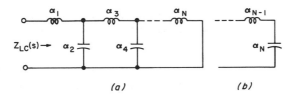

Figure 5.4. A Cauer 1 *LC* network with two possible endings.

the Foster 1 network has exactly N elements, one corresponding to each pole. There is one corresponding to a pole at infinity (the inductance A_∞), one corresponding to a pole at zero (the capacitance $1/A_0$), and two corresponding to each pair of poles at $\pm j\omega_{2k}$ (the inductance A_k/ω_{2k}^2 and the capacitance $1/A_k$). Also, the number of elements N in the Cauer 1 network is equal to the number of poles since there is no premature termination of the continued fraction expansion. This is to be expected, of course, since both networks are general and should have the same number of elements.

The Cauer 2 network is obtained by removing successively the pole at zero, represented by the term A_0/s in (5.3). By an analogous development to that of Cauer 1 we may obtain the continued fraction

$$Z_{LC}(s) = \frac{1}{\beta_1 s} + \cfrac{1}{\cfrac{1}{\beta_2 s} + \cfrac{1}{\cfrac{1}{\beta_3 s} + \cfrac{1}{\ddots \cfrac{1}{\beta_N s}}}} \tag{5.13}$$

resulting in the Cauer 2 network of Fig. 5.5. If Z_{LC} has no pole at zero, then Y_{LC} does and has the form of (5.13.) In this case the first element is a shunt inductor. In either case, Cauer 2 is a ladder network with

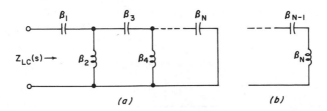

Figure 5.5. A Cauer 2 network with two possible endings.

series capacitors and shunt inductors. By analogy with Cauer 1, the Cauer 2 network may be said to be obtained by an expansion "about zero."

We may note at this point that the ratio of the odd and even parts of a Hurwitz polynomial, as discussed in Sec. 2.2, is an LC function, since it has the form of (5.12) or (5.13). This fact will be very useful later in two-port synthesis.

As an example, consider the function

$$Z_{LC}(s) = \frac{2(s^2 + 1)(s^2 + 3)}{s(s^2 + 2)} \qquad (5.14)$$

which has a pole at infinity. Cauer 1 thus may be found using Z_{LC} rather than Y_{LC}. Noting that (5.12) is obtained as in the Hurwitz test, we may perform the operations in the form

$$s^3 + 2s \,)\, 2s^4 + 8s^2 + 6 \,(\, 2s \quad (\alpha_1 = 2)$$
$$\underline{2s^4 + 4s^2}$$
$$\qquad 4s^2 + 6 \,)\, s^3 + \,\, 2s \,(\, s/4 \quad (\alpha_2 = 1/4)$$
$$\qquad \underline{s^3 + \frac{3}{2}s}$$
$$\qquad\qquad \frac{1}{2}s \,)\, 4s^2 + 6 \,(\, 8s \quad (\alpha_3 = 8) \qquad (5.15)$$
$$\qquad\qquad \underline{4s^2}$$
$$\qquad\qquad\qquad 6 \,)\, \frac{1}{2}s \,(\, s/12 \quad (\alpha_4 = 1/12)$$
$$\qquad\qquad\qquad \underline{\frac{1}{2}s}$$

Therefore Cauer 1 is as shown in Fig. 5.6.

To obtain Cauer 2 we also use Z_{LC}, since in this case it has a pole at the origin. However, we divide with the polynomials in ascending

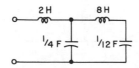

Figure 5.6. A Cauer 1 example.

order, as follows:

$$2s + s^3 \;) \; 6 + 8s^2 + 2s^4 \; (\; 3/s \quad (\beta_1 = 1/3)$$
$$\underline{6 + 3s^2}$$
$$5s^2 + 2s^4 \;) \; 2s + s^3 \quad (\; 2/5s \quad (\beta_2 = 5/2)$$
$$\underline{2s + \frac{4}{5}s^3}$$
$$\frac{1}{5}s^3 \;) \; 5s^2 + 2s^4 \; (\; 25/s \quad (\beta_3 = 1/25)$$
$$\underline{5s^2}$$
$$2s^4 \;) \; \frac{1}{5}s^3 \; (\; 1/10s \quad (\beta_4 = 10)$$
$$\underline{\frac{1}{5}s^3}$$

The Cauer 2 network is shown in Fig. 5.7. The function Z_{LC} has four poles in this case, and both Cauer networks have four elements, as they should.

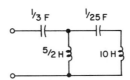

Figure 5.7. A Cauer 2 example.

In these last two sections of this chapter we have obtained the properties of LC driving-point functions that we shall need in two-port synthesis, to be considered next. For a more complete discussion of the LC functions, the reader is referred to such standard works as [V], [W], [Ba], or [Tu].

5.4 SINGLE-TERMINATION REALIZATIONS

One of the simpler passive realization methods is the synthesis of a transfer function by a lossless network terminated in a resistor, as shown in Fig. 5.8. To distinguish this network from the more general one which has, in addition, a resistor at the generator port, we shall refer

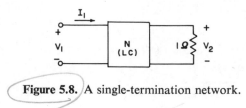

Figure 5.8. A single-termination network.

to Fig. 5.8 as a *single-termination* realization. Because of the concept of network scaling, considered in Sec. 2.6, we may, without loss in generality, take the load resistance as 1Ω. In this case we have, as noted in Sec. 2.3,

$$\frac{V_2}{V_1} = \frac{-y_{21}}{1 + y_{22}} \tag{5.16}$$

and

$$\frac{V_2}{I_1} = \frac{z_{21}}{1 + z_{22}} \tag{5.17}$$

where the y_{ij}'s and z_{ij}'s are the two-port parameters of the LC network N.

In general the transfer function to be realized may be written in the form

$$H(s) = \frac{P(s)}{Q(s)} = \frac{M_1(s) + N_1(s)}{M_2(s) + N_2(s)} \tag{5.18}$$

where P and Q are polynomials, M_1, M_2 are even polynomials, and N_1, N_2 are odd polynomials. It is known [W] that any function of the type in (5.18) is realizable by an LC two-port terminated in a resistor if $Q = M_2 + N_2$ is strictly Hurwitz, and $P(s)$ is an even ($N_1 = 0$) or an odd ($M_1 = 0$) polynomial. In addition, if all zeros of transmission are on the $j\omega$-axis, an LC ladder realization is possible.

In the case P even, the transfer function has the form

$$H(s) = \frac{M_1}{M_2 + N_2} \tag{5.19}$$

and can be put in either of the forms of (5.16) or (5.17) by dividing both numerator and denominator by N_2, resulting in

$$H(s) = \frac{M_1/N_2}{1 + M_2/N_2}$$

This allows us to make the identifications

$$-y_{21} = M_1/N_2, \qquad y_{22} = M_2/N_2 \qquad\qquad (5.20)$$

in (5.16) and

$$z_{21} = M_1/N_2, \qquad z_{22} = M_2/N_2 \qquad\qquad (5.21)$$

in (5.17). The Hurwitz test assures us that M_2/N_2 is a driving-point LC function, as required. This is also true, of course, of N_2/M_2, but the division by N_2 is selected because the transfer function $-y_{21}$ or z_{21} of an LC network must be odd. This requirement is satisfied by M_1/N_2, but would not be by M_1/M_2, which is even.

In the case P odd, the transfer function

$$H(s) = \frac{N_1}{M_2 + N_2} \qquad\qquad (5.22)$$

is put in the form

$$H(s) = \frac{N_1/M_2}{1 + N_2/M_2}$$

so that in (5.16) we have

$$-y_{21} = \frac{N_1}{M_2}, \qquad y_{22} = \frac{N_2}{M_2} \qquad\qquad (5.23)$$

and in (5.17) we have

$$z_{21} = \frac{N_1}{M_2}, \qquad z_{22} = \frac{N_2}{M_2} \qquad\qquad (5.24)$$

Let us consider first the case where $Q = M_2 + N_2$ is strictly Hurwitz of degree n and $P = Ks^m$, for $0 \le m \le n$, and K a real constant. This is the special case where the transmission zeros are all at zero and/or infinity, and is very important since it includes, as we have seen, low-pass ($m = 0$), high-pass ($m = n$), and bandpass ($m = n/2$, n even) filter functions.

We shall restrict our attention to the voltage-ratio transfer function (the details in the other cases are virtually the same, as seen by (2.18) and (2.19)). Let us begin with the case $m = 0$,

$$H(s) = \frac{V_2}{V_1} = \frac{K}{M_2 + N_2} \qquad\qquad (5.25)$$

which is an all-pole function. Since the numerator is even, we have by (5.20)

$$-y_{21} = \frac{K}{N_2(s)}, \qquad y_{22} = \frac{M_2(s)}{N_2(s)} \qquad (5.26)$$

To realize an LC two-port network with parameters given by (5.26) and satisfying (5.25) when terminated with a 1Ω resistor, we make the following observations. Both y_{22} and $-y_{21}$ have the same denominator and all the zeros of $-y_{21}$ are at infinity. Thus an LC network with all its zeros of transmission at infinity, which realizes y_{22} exactly, will realize $-y_{21}$ within a constant multiplier. For both functions will have the same denominator and the numerator of $-y_{21}$ will be a constant, since all its zeros will be at infinity. Of the several ways we have of realizing the driving-point function y_{22}, such as the Foster and Cauer networks, are there any with all zeros of transmission at infinity?

To answer this question, let us consider the ladder network, a general form of which is shown in Fig. 5.9. If an input is applied at

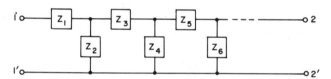

Figure 5.9. A general ladder network.

terminals 1–1' at a frequency s_1, a zero of transmission, then there is no output at terminals 2–2'. In a ladder the only way there can be no transmission (no output) is for one of the series arms to be an open-circuit (one of the impedances $Z_1, Z_3, Z_5, \ldots$, to be infinite), or for one of the shunt arms to be a short-circuit (one of the impedances $Z_2, Z_4, Z_6, \ldots$, to be zero) at the frequency s_1. In the former case the current is blocked and in the latter case it is diverted through a short-circuit.

To return to our problem, there is one realization, the Cauer 1, of y_{22} which has all its zeros of transmission at infinity. Its series arms are inductors, which become open-circuits at infinite frequency, and its shunt arms are capacitors, which become short-circuits at infinite frequency. Therefore to realize (5.26), we have only to realize y_{22} in a Cauer 1 network looking in the 2–2' terminals. We must be careful to terminate the network at the 1–1′ terminals so that when they are shorted we see y_{22} from the 2–2′ terminals.

As an example, consider the third-order Butterworth function,

$$\frac{V_2}{V_1} = \frac{K}{s^3 + 2s^2 + 2s + 1} \tag{5.27}$$

which is equivalent to

$$\frac{V_2}{V_1} = \frac{\dfrac{K}{s^3 + 2s}}{1 + \dfrac{2s^2 + 1}{s^3 + 2s}}$$

Therefore we have

$$-y_{21} = \frac{K}{s^3 + 2s}, \qquad y_{22} = \frac{2s^2 + 1}{s^3 + 2s} \tag{5.28}$$

The Cauer 1 expansion for y_{22} is given by

$$2s^2 + 1) \; s^3 + \quad 2s \; (\; s/2 \longleftarrow Z$$

$$\underline{s^3 + \frac{1}{2}s}$$

$$\frac{3}{2}s \;) \; 2s^2 + 1 \; (\; 4s/3 \longleftarrow Y$$

$$\underline{2s^2}$$

$$1 \;) \; 3s/2 \; (\; 3s/2 \longleftarrow Z$$

$$\underline{\underline{3s/2}}$$

and since y_{22} was inverted to obtain the pole at infinity, the first element, looking in the 2–2′ port, is a series impedance of $s/2$. Next is a shunt capacitance, etc. The circuit is shown in Fig. 5.10, with the 1Ω termination. Analysis shows that (5.27) is realized with $K = 1$.

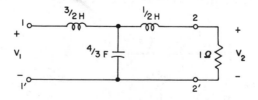

Figure 5.10. A realization of Eq. (5.27) with $K = 1$.

Let us consider next the case $m = n$, where

$$H(s) = \frac{Ks^n}{M_2(s) + N_2(s)} = \frac{Ks^n}{s^n + a_{n-1}s^{n-1} + \cdots + a_1 s + a_0} \quad (5.29)$$

In this case all the zeros of transmission are at zero. The Cauer 2 network has this feature, since the series arms are capacitors which are open-circuits at $s = 0$, and the shunt arms are inductors which are short-circuits at $s = 0$. Therefore we may synthesize (5.29) by obtaining the two-port parameters as described earlier and realizing the driving-point parameter with a Cauer 2 network.

Finally, let us consider the case $0 < m < n$, for which $H(s)$ has some of its zeros at zero and the rest at infinity. From the earlier discussion it should be clear that $H(s)$ can be synthesized by realizing the driving-point function y_{22} or z_{22} with a mixture of the Cauer 1 and Cauer 2 circuits. That is, we begin as in the Cauer 1 (or Cauer 2) and obtain the terms in the expansion corresponding to transmission zeros at infinity (or zero). Then when all these zeros are obtained, switch to the other type Cauer network to complete the process.

As an example, suppose we want to realize

$$\frac{V_2}{V_1} = \frac{Ks^3}{s^4 + s^3 + 4s^2 + 2s + 3}$$

$$= \frac{\dfrac{Ks^3}{s^4 + 4s^2 + 3}}{1 + \dfrac{s^3 + 2s}{s^4 + 4s^2 + 3}} \quad (5.30)$$

The driving-point function

$$y_{22} = \frac{s^3 + 2s}{s^4 + 4s^2 + 3}$$

is to be realized by a network with three transmission zeros at zero and

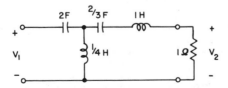

Figure 5.11. A realization of Eq. (5.30) with $K = 1$.

$n - m = 4 - 3 = 1$

one at infinity, as we see from (5.30). We shall start with a Cauer 1 expansion, obtain one zero at infinity, then switch to a Cauer 2 expansion and obtain the three zeros at zero. (We could start as well with Cauer 2.) The expansion is given by

$$
\begin{array}{r}
s^3 + 2s \ )\ s^4 +\ \ 4s^2 + 3 \quad (\ s \longleftarrow Z \\
\underline{s^4 +\ \ 2s^2} \\
2s^2 + 3 \quad (\ 3/2s \longleftarrow Z \\
\underline{\tfrac{3}{2}s^2 + 3} \\
\tfrac{1}{2}s^2\ )\ s^3 +\ \ 2s \quad (\ 4/s \longleftarrow Y \\
\underline{2s} \\
s^3\ )\ \tfrac{1}{2}s^2\ (\ 1/2s \longleftarrow Z \\
\underline{\underline{\tfrac{1}{2}s^2}}
\end{array}
$$

The network is shown in Fig. 5.11, with the required 1Ω termination. Analysis shows it to yield (5.30) with $K = 1$.

We note in this example that the continued fraction expansion required that the first two terms be obtained without inverting, so that two consecutive Zs were obtained. This was interpreted as two series elements in the network.

5.5 ZERO SHIFTING

In the previous section, all the transmission zeros were at zero and/or infinity, and the realization of the driving-point parameter in a Cauer 1, Cauer 2, or combination of Cauer 1 and Cauer 2 network automatically realized the transfer parameter. This is because in the continued fraction expansion process the zeros of transmission were automatically *shifted* to the appropriate places, namely zero and/or infinity. In the next section we shall consider the same type of functions as in the previous section except that some of the zeros are on the finite, nonzero portion of the $j\omega$-axis. In these cases the *zero shifting* process will have to be done deliberately, since it will not be automatic.

In the Cauer ladder development we are essentially removing one pole at a time, inverting the remainder after each removal, and repeating the process. If the term involved is a Z_1 with a zero at $j\omega_1$, then the development of the ladder requires that $Y_1 = 1/Z_1$ be realized as the next element. Then Y_1 will have a pole at $j\omega_1$, and will appear as a shunt element, which will be a short-circuit at $s = j\omega_1$. Thus the zero of Z_1 is a zero of transmission. If on the other hand, the term involved at some stage is a Y_2 with a zero at $s = j\omega_1$, then it will be realized next as $Z_2 = 1/Y_2$ in a series arm. This series arm will then be an open-circuit at $s = j\omega_1$, and thus the zero of Y_2 is a zero of transmission. In summary, in the ladder development it does not matter whether the function with the zero $j\omega_1$ at any stage is a Z or a Y; inverting the function and removing the resulting pole at $j\omega_1$ realizes a zero of transmission at $j\omega_1$.

Our problem then in realizing zeros of transmission is to realize the driving-point function exactly, in a manner such that each transmission zero will eventually appear in a function which is to be synthesized as a ladder element. In the previous section the zeros appeared this way as a matter of course, but in the more general case we need a means of placing the zeros where we want them.

Let us consider first the idea of pole removals to realize a network element. If $Z_{LC}(s)$ has a pole at infinity, then by (5.3) we may write

$$Z_{LC}(s) = A_\infty s + Z_1(s) \tag{5.31}$$

where

$$Z_1(s) = \frac{A_0}{s} + \sum_{k=1}^{n} \frac{A_k s}{s^2 + \omega_{2k}^2} \tag{5.32}$$

We may partially realize Z_{LC} by interpreting (5.31) as an inductance of A_∞ in series with Z_1. In this case the pole at infinity has been removed since Z_1 has no pole at infinity.

Now let us consider the case

$$Z_{LC}(s) = k_p A_\infty s + Z_1(s) \tag{5.33}$$

where $0 < k_p < 1$, and

$$Z_1(s) = (1 - k_p)A_\infty s + \frac{A_0}{s} + \sum_{k=1}^{n} \frac{A_k s}{s^2 + \omega_{2k}^2}$$

As before, we interpret (5.33) as an inductance $k_p A_\infty$ in series with Z_1, but in this case the pole at infinity has not been removed, since it is still contained in Z_1. This operation is called "partially" removing the pole at infinity. Similar remarks may be made concerning removal and partial removal of poles at zero and at $\pm j\omega_1, 0 < \omega_1 < \infty$. The concept of partial removal of poles plays a prominent role, as we shall see, in the zero-shifting process required in realizing finite, nonzero transmission zeros.

As we have seen in Sec. 5.2, the reactance $X(\omega)$ of $Z_{LC}(s)$ has a positive slope for all ω. A plot of $X(\omega)$ with poles at zero and infinity is shown in Fig. 5.12, which we shall use to illustrate how zero shifting can be accomplished by removal or partial removal of a pole at infinity.

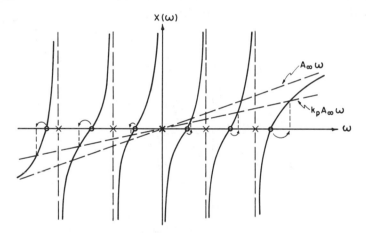

Figure 5.12. A plot of $X(\omega)$, illustrating the effect of partial removal of a pole at infinity.

We note that as ω becomes infinite, $X(\omega) \rightarrow A_\infty\omega$, from (5.31) and (5.32). This is shown by the asymptote labeled $A_\infty\omega$ in Fig. 5.12.

Let us partially remove the pole at infinity in accordance with (5.33). Then for $s = j\omega$ we have

$$X(\omega) = k_p A_\infty\omega + X_1(\omega)$$

Therefore $X_1(\omega) = 0$ when $X(\omega) = k_p A_\infty\omega$, which occurs at the points of intersection of $X(\omega)$, represented by the solid line, and $k_p A_\infty\omega$, represented by the broken line, as shown in Fig. 5.12. Thus the zeros of the remainder $Z_1(s)$ are those of $Z(s)$ shifted as indicated by the arrows. In every case the shift is toward infinity. We note also that if the pole at infinity is completely removed ($k_p = 1$), then the new zeros are at the intersection of $X(\omega)$ and $A_\infty\omega$, indicating that the largest zero shifts all the way to infinity. This is the case in the Cauer 1 development, and accounts for the automatic shift of the zeros to the proper position.

If we partially remove the pole at zero, we have, for $0 < k_p < 1$,

$$X(\omega) = -\frac{k_p A_0}{\omega} + X_1(\omega) \tag{5.34}$$

where $X_1(\omega)$ is the reactance of the remainder term. The zeros of $X_1(\omega)$ occur when $X(\omega) = -k_p A_0/\omega$ and, as may be seen from Fig. 5.13, are the zeros of $X(\omega)$ shifted toward zero.

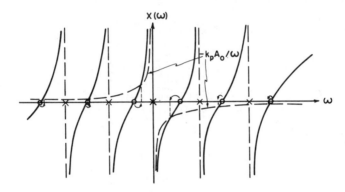

Figure 5.13. A plot illustrating the effect of partial removal of a pole at zero.

Finally, partially removing the pole at $\pm j\omega_1$ (ω_1 finite and nonzero) results in

$$X(\omega) = \frac{k_p A_1 \omega}{-\omega^2 + \omega_1^2} + X_1(\omega)$$

where $0 < k_p < 1$ and $X_1(\omega)$ is the reactance of the remainder $Z_1(s)$. The zeros of $X_1(\omega)$ are those of $X(\omega)$ shifted toward ω_1, as may be seen from Fig. 5.14.

In each case the zeros have shifted toward the pole which is partially removed, whether it be at zero, infinity, or at $j\omega_1$. We have shown this for the special cases of $X(\omega)$ in Figs. 5.12, 5.13, and 5.14; however, it is

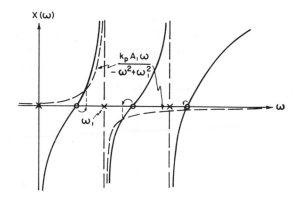

Figure 5.14. A plot illustrating the effect of partial removal of a pole at $j\omega_1$.

true in general, being a consequence of the positive slope of $X(\omega)$. In the next section we shall use the zero-shifting technique to realize more general two-port functions.

5.6   REALIZATIONS REQUIRING ZERO SHIFTING

We are now ready to consider the synthesis of more general network functions having a Hurwitz denominator and an odd or even numerator with all its zeros on the $j\omega$-axis. In this case we shall be interested in realizing a two-port driving-point parameter such as y_{22} so as to obtain transmission zeros at finite nonzero points on the $j\omega$-axis, as well as at possibly the origin and infinity. Zeros at infinity or the origin may be realized by steps in a Cauer 1 or Cauer 2 continued fraction expansion. The other zeros may be realized by zero shifts using partial removal of poles, as described in the previous section.

As an example, let us consider the transfer function

$$\frac{V_2}{V_1} = \frac{Ks(s^2 + 1)}{s^3 + 2s^2 + 2s + 1} \tag{5.35}$$

which has zeros at $s = 0$, $s = \pm j1$. Dividing numerator and denominator by $2s^2 + 1$ yields

$$-y_{21} = \frac{Ks(s^2 + 1)}{2s^2 + 1}, \qquad y_{22} = \frac{s(s^2 + 2)}{2s^2 + 1} \tag{5.36}$$

Since the ladder development entails zero shifting, inverting, and realizing elements by removing poles, we shall find it convenient to keep track of the process by plotting the poles and zeros for the various functions as we proceed through the ladder. This is done in Fig. 5.15, where the first two plots are the pole-zero plots of $-y_{21}$ and y_{22}.

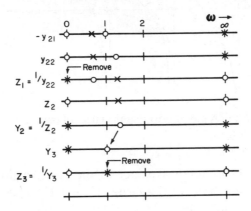

Figure 5.15. Pole-zero plots of the various functions involved in the ladder development.

We note that y_{22} has a zero at zero, as does $-y_{21}$, so that writing

$$Z_1 = \frac{1}{y_{22}} = \frac{2s^2 + 1}{s(s^2 + 2)}$$

we may remove the pole at the origin and hence realize the transmission zero at $s = 0$. The pole-zero plot of Z_1 is shown in Fig. 5.15, with the pole at $s = 0$ specified for removal.

Thus we have

$$Z_1 = \frac{1}{2s} + Z_2$$

where

$$Z_2 = \frac{\frac{3}{2}s}{s^2 + 2}$$

At this point we have partially realized y_{22} as shown in Fig. 5.16; also the pole-zero plots of Z_2 and its reciprocal

$$Y_2 = 1/Z_2 = \frac{2(s^2 + 2)}{3s}$$

are shown in Fig. 5.15.

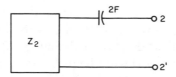

Figure 5.16. A partial realization of y_{22}.

We need to shift a zero to $\omega = 1$ ($s = j1$) in order to realize the remaining zero of transmission. (Note that realization of $j1$ automatically realizes its conjugate $-j1$, the other transmission zero.) From Sec. 5.5 we recall that the zero at $j\sqrt{2}$ of Y_2 can be shifted toward zero by partial removal of the pole at $s = 0$. Thus we write

$$Y_2 = \frac{k_p A_0}{s} + Y_3$$

where we require $Y_3(j1) = 0$, or

$$Y_2(j1) = \frac{2(-1+2)}{j3} = \frac{k_p A_0}{j1}$$

Thus we have $k_p A_0 = \frac{2}{3}$, or

$$Y_2 = \frac{2}{3s} + Y_3 \tag{5.37}$$

and hence

$$Y_3 = Y_2 - \frac{2}{3s} = \frac{2(s^2 + 1)}{3s}$$

which has the required zero at $j1$. Pole-zero plots of Y_3 and its reciprocal

$$Z_3 = \frac{1}{Y_3} = \frac{3s}{2(s^2 + 1)} \tag{5.38}$$

are shown in Fig. 5.15. The network development at this stage, obtained from Fig. 5.16 and (5.37), is shown in Fig. 5.17.

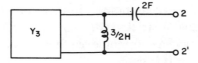

Figure 5.17. A further development of y_{22}.

Finally, removing the pole of Z_3 at $j1$ (and simultaneously removing the pole at $-j1$) completes the synthesis, as the final pole-zero plot of Fig. 5.15 indicates. This is done by realizing (5.38) and completing Fig. 5.17. The completed two-port with the required 1Ω termination is shown in Fig. 5.18. Analysis shows that (5.35) is realized with $K = 1$.

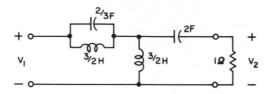

Figure 5.18. A realization of Eq. (5.35).

5.7 PRIVATE POLES

In the realizations considered thus far in this chapter it has been assumed that the transfer parameter $(-y_{21}$ or z_{21}, say) had the same poles as the driving-point parameter $(y_{22}$ or z_{22}, say). It may happen because of cancellation that the driving-point parameter has *private poles*, that is, poles not shared by the transfer parameter. For example, consider the transfer function,

$$\frac{V_2}{V_1} = \frac{K(s^2 + 2)}{s^3 + 2s^2 + 2s + 1}$$

$$= \frac{-y_{21}}{1 + y_{22}} \tag{5.39}$$

where

$$y_{22} = \frac{2s^2 + 1}{s(s^2 + 2)}$$

$$-y_{21} = \frac{K(s^2 + 2)}{s(s^2 + 2)} = \frac{K}{s} \tag{5.40}$$

The function y_{22} has a pole at zero and the private poles, $s = \pm j\sqrt{2}$.

It is also possible for the transfer function to be of a form which results in cancellation in the driving-point parameter, so that the transfer parameter has private poles. However, in this case the transfer function does not have a strictly Hurwitz denominator, since its even and odd parts must have a common even factor. As stated at the outset, we are not considering this case.

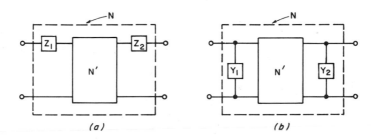

Figure 5.19. Two-port networks resulting in private poles.

If the driving-point parameter has private poles, the two-port synthesis is easily effected, as we may see from Figs. 5.19(a) and (b). Let z_{ij} and z'_{ij}, $i, j = 1, 2$, be respectively the open-circuit impedances of the two-ports N and N' of Fig. 5.19(a). Then evidently we have

$$z_{11} = Z_1 + z'_{11}$$
$$z_{12} = z'_{12}, z_{21} = z'_{21} \qquad (5.41)$$
$$z_{22} = Z_2 + z'_{22}$$

Thus any private poles of z_{11} (or of z_{22}) may be removed in a series impedance Z_1 (or Z_2). Then the remainder z'_{11} (or z'_{22}) has the same poles as the transfer impedance and the process may be continued as in the previous sections of this chapter.

A dual argument holds if y_{11} (or y_{22}) has private poles, since from Fig. 5.19(b) we have

$$y_{11} = Y_1 + y'_{11}$$
$$y_{12} = y'_{12}, y_{21} = y'_{21} \qquad (5.42)$$
$$y_{22} = Y_2 + y'_{22}$$

Thus private poles of y_{11} (or y_{22}) may be removed in a shunt admittance Y_1 (or Y_2), and the remainder function y'_{11} (or y'_{22}) has the same poles as the transfer admittance.

As an example, let us realize the functions of (5.40). We have

$$y_{22} = \frac{2s^2 + 1}{s(s^2 + 2)} = \frac{\frac{3}{2}s}{s^2 + 2} + \frac{1/2}{s}$$

so that, referring to Fig. 5.19(b),

$$Y_2 = \frac{\frac{3}{2}s}{s^2 + 2}$$

and N' has parameters

$$y'_{22} = \frac{1/2}{s}, \qquad -y'_{21} = \frac{K}{s}$$

The resulting realization is shown in Fig. 5.20. The transfer function (5.39) is realized for $K = \frac{1}{2}$ by loading the network with the 1Ω termination, as shown.

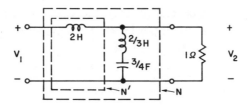

Figure 5.20. A realization of Eq. (5.39).

5.8 SUMMARY

In this chapter the properties and methods of passive synthesis of driving-point LC networks were considered and extended to passive realizations of two-port network functions. The two-port networks obtained were of the LC ladder type with either an open secondary or a secondary terminated in a 1-ohm resistor. The elements in the LC ladder may be obtained directly as Cauer-type networks, or by means of a zero-shifting procedure, required when the transmission zeros are not all at zero and/or infinity. More general network structures, such as lattices and doubly-terminated ladders will be considered in Chapters 7 and 9.

EXERCISES

5.1. Show that (a) $P(s) = s^4 + a$, $a > 0$, and (b) $P(s) = s^4 + (2\beta - \alpha)s^2 + \beta^2$, $2\beta \geq \alpha > 0$, are not Hurwitz, and their zeros constitute a quad of zeros.

5.2. Given $Z(s) = \dfrac{(s^2 + 1)(s^2 + 9)}{s(s^2 + 4)}$.

Find a Foster 1, Cauer 1, and Cauer 2 realization. Plot $X(\omega)$ versus ω for $0 \le \omega \le 4$.

5.3. Observe that $Y_{LC}(s)$ and $Z_{LC}(s)$ are identical in form and therefore $Y_{LC}(s)$ has the form

$$Y_{LC}(s) = B_\infty s + \frac{B_0}{s} + \sum_{k=1}^{m} \frac{B_k s}{s^2 + \omega_{2k-1}^2}$$

(a) Interpret this in a general network to obtain the Foster 2 realization. (b) Find the Foster 2 network for the function of Exercise 5.2. Answer (a):

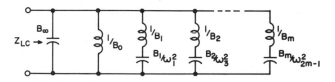

Figure Ex. 5.3.

5.4. Show that the number of elements in each of the networks, Foster 1, Foster 2, Cauer 1, and Cauer 2, of a Z_{LC} is equal to the number of poles of Z_{LC}. Show also that each network has the same number of Ls and of Cs.

5.5. Obtain an LC two-port network terminated in a 1Ω resistor having

$$\frac{V_2}{V_1} = \frac{P(s)}{s^4 + s^3 + 34s^2 + 16s + 225}$$

where $P(s)$ is given by (a) K, (b) Ks, (c) Ks^2, and (d) Ks^4.

5.6. Find a realization like that of Exercise 5.5 for the high-pass Butterworth filter function

$$H(s) = \frac{V_2}{V_1} = \frac{Ks^3}{s^3 + 2s^2 + 2s + 1}$$

Repeat the exercise for $H(s) = V_2/I_1$.

5.7. Note that in Exercise 5.6, for $s \to \infty$, we have $H(s) \to K$. Therefore by analyzing the circuit for $s \to \infty$, compare the results and determine K.

5.8. Note that in (5.3) we may find A_k by the formula

$$A_k - \left(\frac{s^2 + \omega_{2k}^2}{s}\right) Z_{LC}(s) \Big|_{s^2 = -\omega_{2k}^2}$$

Use this to obtain a Foster 1 realization of

$$Z_{LC}(s) = \frac{s(s^2 + 2)(s^2 + 4)}{(s^2 + 1)(s^2 + 3)(s^2 + 5)}$$

5.9. (a) Show that if $z_{12} = z_{21}$, a general T equivalent network of a 3 terminal 2-port network with specified z-parameters is given as shown in Fig. Ex. 5.9.

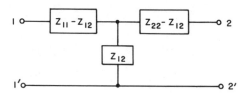

Figure Ex. 5.9.

(b) Since any 3-terminal RLC network can be transformed to a T by Y-Δ or star-mesh transformations, the expressions for the three network impedances cannot contain negative coefficients. If z_{11}, z_{12}, and z_{22} all have the same denominators, show that the coefficients of s^k in the numerator of z_{12} cannot exceed the coefficients of s^k in either the numerator of z_{11} or of z_{22}.

(c) What can be concluded about the upper limit on K in Exercise 5.5 from (b)?

(d) What are the limitations on K in the general case?

5.10. Realize the RLC driving-point function

$$Y(s) = \frac{s(2s^2 + s + 2)}{(s^2 + 1)(s^2 + s + 1)}$$

by first removing the poles at $\pm j1$.

5.11. Repeat Exercise 5.5 for the case $P(s) = K(s^4 + 5s^2 + 4)$

5.12. Synthesize as an LC ladder terminated in a 1Ω resistor, the Butterworth function

$$\left| \frac{V_2(j\omega)}{V_1(j\omega)} \right|^2 = \frac{K}{1 + \left(\frac{\omega}{1,000}\right)^4}$$

Scale the resulting network so that the only capacitors used are $1\mu F$.

5.13. Obtain a Butterworth bandpass filter of order 4 with $\omega_0 = 10,000$ rad/s and $Q = 10$, using a single termination resistance of $1k\Omega$.

5.14. Obtain $H(s) = V_2(s)/V_1(s)$ for a sixth-order Butterworth bandpass filter with $\omega_0 = 1$ rad/s and $Q = 5$. Realize $H(s)$ with an LC two-port terminated in 1Ω and a center frequency of $\omega_0 = 10{,}000$ rad/s. Impedance scale the result to obtain elements as practical as possible.

5.15. Consider the following filter functions with ω_c or ω_0 normalized to 1 rad/s. Obtain single-termination networks with ω_c or ω_0 denormalized to 10,000 rad/s and terminations of $1k\Omega$:

(a) $\dfrac{V_2}{V_1} = \dfrac{K}{s^3 + s^2 + 1.25s + 0.5}$ (low-pass Chebyshev, $\epsilon = 0.5$)

(b) $\dfrac{V_2}{V_1} = \dfrac{Ks^3}{s^3 + 2.5s^2 + 2s + 2}$ (high-pass Chebyshev, $\epsilon = 0.5$)

5.16. Using the Chebyshev function of Exercise 5.15(a) as the low-pass prototype, obtain V_2/V_1 for a 6th-order bandpass filter, and obtain a single-termination realization with $Q = 10$, $\omega_0 = 10{,}000$ rad/s, and the termination of $1k\Omega$.

5.17. Repeat Exercise 5.16 for a band-reject filter, rather than a band-pass filter.

5.18. Note that the low-pass to bandpass transformation of (4.15) may be interpreted as $S = Z_{LC}(s)$, where $Z_{LC}(s)$ has a pole at zero and at infinity. The mappings $S = j1$ onto $s = j\omega_U$ and $S = -j1$ onto $s = j\omega_L$, etc., are possible because of the nature of the plot of $X(\Omega)$ versus Ω. Extend this idea to the case of two separate passbands ($S = Z_{LC}(s)$ where $Z_{LC}(s)$ is a quartic over a cubic), and in general n separate passbands.

5.19. The driving-point impedance $Z_{RC}(s)$ of a general RC one-port may be obtained using (2.4) and (2.2), which become in this case,

$$Z_{RC}(s) = \frac{\Delta(s)}{\Delta_{11}(s)}, \qquad Z_{ij}^{RC}(s) = R_{ij} + \frac{C_{ij}^{-1}}{s}$$

Show that

$$pZ_{ij}^{RC}(p^2) = Z_{ij}^{LC}(p)$$

and hence that *Cauer's transformation*,

$$Z_{RC}(s) = \frac{1}{p}Z_{LC}(p)\bigg|_{p^2 = s}$$

is valid.

5.20. Using the results of Exercise 5.19 and Eq. (5.1), show that the general RC impedance may be written

$$Z_{RC}(s) = \frac{K(s + \sigma_1)(s + \sigma_3) \cdots (s + \sigma_{2m-1})}{s(s + \sigma_2)(s + \sigma_4) \cdots (s + \sigma_{2n})}$$

where $K > 0$, $m = n$ or $n + 1$, and $0 \le \sigma_1 < \sigma_2 < \sigma_3 < \cdots$. Thus the poles and zeros interlace on the negative real axis, and the pole or zero nearest or at the origin is a pole.

5.21. Using Exercise 5.19 and Eq. (5.3), show that in general

$$Z_{RC}(s) = A_\infty + \frac{A_0}{s} + \sum_{k=1}^{n} \frac{A_k}{s + \sigma_{2k}}$$

and interpret this result to obtain the *Foster* 1 *RC* network shown in Fig. Ex. 5.21.

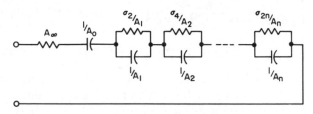

Figure Ex. 5.21.

5.22. Using Exercise 5.19 and Eq. (5.12), show that

$$Z_{RC}(s) = \alpha_1 + \cfrac{1}{\alpha_2 s + \cfrac{1}{\alpha_3 + \cfrac{1}{\alpha_4 s + \cdots}}}$$

and from this result find the *Cauer* 1 *RC* network.

5.23. Obtain the form of $Y_{RC}(s)$ in general, using Exercise 5.19, and find the *Foster* 2 and *Cauer* 2 *RC* networks. (Suggestion: Show that $Y_{RC}(s)/s$ has the same form as $Z_{RC}(s)$.)

5.24. Show that $Z_{RL}(s)$, the driving-point impedance of the general *RL* one-port network, has the same form as $Y_{RC}(s)$, and consequently $Y_{RL}(s)$ looks like $Z_{RC}(s)$. Use these results to find the Foster 1, Foster 2, Cauer 1, and Cauer 2 *RL* networks. (*Suggestion:* Consider Exercises 5.19 through 5.23.)

5.25. Find the four networks of Exercise 5.24 for the function

$$F(s) = \frac{(s + 1)(s + 3)}{s(s + 2)}$$

interpreted (a) as a $Z(s)$, and (b) as a $Y(s)$.

5.26. Synthesize with an *LC* two-port terminated in a 1Ω resistor, the function,

$$H(s) = \frac{Ks(s^2 + 1)}{s^4 + s^3 + 4s^2 + 2s + 3}$$

where (a) $H(s) = V_2(s)/V_1(s)$ and (b) $H(s) = V_2(s)/I_1(s)$.

CHAPTER

6

Approximation— Rational Transfer Functions

6.1 INVERSE CHEBYSHEV FILTERS

The Butterworth and Chebyshev filters considered earlier in Chapter 3 are all-pole filters. If the function $f(\omega^2)$ in (3.1) is a rational function with finite poles, as has been noted earlier the transfer function is a rational function with finite zeros. In this chapter we shall consider two such filter functions, the inverse Chebyshev and the elliptic filter functions.

A filter whose transfer function has finite zeros and which exhibits a flat passband response and a stopband with ripples, is the so-called *inverse Chebyshev filter*, which we consider in this section.

If the Chebyshev function

$$|H_c(j\omega)|^2 = \frac{1}{1 + \epsilon^2 C_n^2(\omega)} \tag{6.1}$$

discussed in Chapter 3, is subtracted from 1, then the resulting function is that of a high-pass filter with ripples in the stopband $0 < \omega < 1$. The function $|H_c(j\omega)|^2$ is shown in Fig. 6.1(a) and $1 - |H_c(j\omega)|^2$ is shown in Fig. 6.1(b).

If we now perform a low-pass to high-pass transformation by replacing ω by $1/\omega$, we have the function

$$|H(j\omega)|^2 = 1 - |H_c(j/\omega)|^2 \tag{6.2}$$

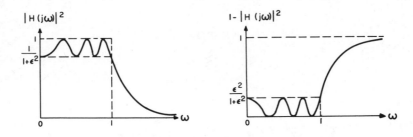

Figure 6.1. (a) A Chebyshev response $|H_C(j\omega)|^2$, and (b) $1 - |H_C(j\omega)|^2$.

which is sketched in Fig. 6.2. The function $|H(j\omega)|$ is the inverse Chebyshev amplitude function and is evidently a low-pass function. We note from Fig. 6.2 that $\omega = 1$ is the beginning of the stopband ripple channel, and not the cutoff point. This is true in the general case, as is clear from the steps in the development.

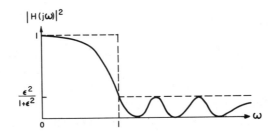

Figure 6.2. $|H(j\omega)|^2$ for an inverse Chebyshev filter.

From (6.1) and (6.2) we obtain the inverse Chebyshev function

$$|H(j\omega)|^2 = \frac{\epsilon^2 C_n^2(1/\omega)}{1 + \epsilon^2 C_n^2(1/\omega)} \tag{6.3}$$

The transfer function of the inverse Chebyshev filter may be found in the usual way from (6.3) with $\omega^2 = -s^2$, etc. It will be a ratio of two polynomials,

$$H(s) = \frac{P(s)}{Q(s)} \tag{6.4}$$

where $P(s)$ will have the finite zeros of the stopband response. The polynomial $Q(s)$ may be found readily for a given n and ϵ by first finding its counterpart $Q_c(s)$ for the Chebyshev filter; this was outlined in Sec. 3.8. Then if $Q_c(s)$ is of degree n, we have

$$Q(s) = s^n Q_c(1/s) \tag{6.5}$$

This may be seen from the fact that the amplitude functions which yield $Q(s)$ and $Q_c(s)$ are identical except that ω is replaced by $1/\omega$ in one of them.

The numerator polynomial $P(s)$ is found easily from

$$P(s)P(-s) = \omega^{2n} C_n^2(1/\omega)\Big|_{\omega^2 = -s^2} \tag{6.6}$$

Since $P(s)P(-s)$ will contain only factors like $(s^2 + a)^2$ (because C_n is either even or odd with real zeros), both $P(s)$ and $P(-s)$ will contain the factor $s^2 + a$.

As an example, suppose we want $H(s)$ for an inverse Chebyshev filter with $n = 3$ and $\epsilon = 0.5$. These are the same values used in the example of the Chebyshev response of Sec. 3.8, which led to

$$Q_c(s) = s^3 + s^2 + 1.25s + 0.5$$

Thus we have

$$Q(s) = s^3 \left(\frac{1}{s^3} + \frac{1}{s^2} + \frac{1.25}{s} + 0.5 \right)$$
$$= 0.5(s^3 + 2.5s^2 + 2s + 2)$$

By (6.6) we have

$$P(s)P(-s) = \omega^6 \left(\frac{4}{\omega^3} - \frac{3}{\omega} \right)^2 \Big|_{\omega^2 = -s^2}$$
$$= (4 + 3s^2)^2$$

or

$$P(s) = 3s^2 + 4$$

Finally, the transfer function is given by

$$H(s) = \frac{K(3s^2 + 4)}{s^3 + 2.5s^2 + 2s + 2}$$

where we have disregarded the factor 0.5 in $Q(s)$ and the factor ϵ^2 in (6.3), lumping them in with the coefficient $3K$ of s^2 in the numerator.

If we wish $H(0) = 1$, as shown in Fig. 6.2, we may choose $K = 0.5$. In this case we may check our work by computing

$$|H(j\omega)|^2 = \frac{(0.5)^2(4 - 3\omega^2)^2}{(2\omega - \omega^3)^2 + (2 - 2.5\omega^2)^2}$$

at $\omega = 1$, resulting in

$$|H(j1)|^2 = \frac{(0.5)^2}{1 + (0.5)^2} = \frac{\epsilon^2}{1 + \epsilon^2}$$

As has been noted, the inverse Chebyshev filter does not have its cutoff at $\omega = 1$. It may be easily shown that the conventional 3 dB cutoff point is given by

$$\omega_c = \frac{1}{\cosh\left(\frac{1}{n} \cosh^{-1} \frac{1}{\epsilon}\right)} \tag{6.7}$$

(The reader is asked to obtain this result in Exercise 6.1.)

The inverse Chebyshev filter has finite stopband zeros, but its flat passband characteristics cancel any improvement which finite zeros contribute to the attenuation rate. Consequently a Chebyshev filter of the same order will satisfy the same specifications [H], [W], and since the synthesis of the inverse Chebyshev filter requires zero-shifting, and hence more elements, it is rarely used instead of the Chebyshev. It is, however, more economical in elements than a Butterworth filter meeting the same requirements [W]. There may be cases also when time-domain considerations make the inverse Chebyshev filter preferable to the Chebyshev filter, as we shall see in Chapter 8.

6.2 ELLIPTIC FILTERS

Thus far we have considered low-pass amplitude functions which are monotonic everywhere (Butterworth), monotonic in the stopband with passband ripples (Chebyshev), or monotonic in the passband with stopband ripples (inverse Chebyshev). In the first two cases $f(\omega^2)$ in (3.1) is a polynomial, which results in an all-pole transfer function. In the case of the inverse Chebyshev, $f(\omega^2)$ is a rational function with all its zeros at $\omega = 0$ and its poles in the stopband, accounting for the stopband transmission zeros.

In the most general case $f(\omega^2)$ is a rational function with poles and zeros which are finite and/or infinite. At the zeros of $f(\omega^2)$, $|H(j\omega)|$ attains its maximum value of 1, and at the poles of $f(\omega^2)$, $|H(j\omega)| = 0$. Thus by properly choosing the numerator and denominator of $f(\omega^2)$ so that its zeros are all in the passband and its poles are all in the stop-band, we may obtain a very good approximation to the ideal low-pass case. Evidently in the case where $f(\omega^2)$ has its zeros distributed across the passband (rather than all concentrated at one point) and its poles distributed across the stopband, the amplitude function will exhibit ripples in both the pass- and stopbands.

One way to assure that the poles of $f(\omega^2)$ are in the stopband when its zeros are in the passband is to choose the poles as the reciprocals of the zeros. This may be done by letting

$$f(\omega^2) = \epsilon^2 R_n^2(\omega) \tag{6.8}$$

where, since R_n may be either even or odd,

$$R_n(\omega) = \prod_{i=1}^{[n/2]} \frac{(\omega_{2i-1}^2 - \omega^2)}{(1 - \omega_{2i-1}^2\omega^2)}; \qquad n = 2, 4, 6, \ldots \tag{6.9}$$

or

$$R_n(\omega) = \omega \prod_{i=1}^{[n/2]} \frac{(\omega_{2i}^2 - \omega^2)}{(1 - \omega_{2i}^2\omega^2)}; \qquad n = 3, 5, 7, \ldots \tag{6.10}$$

The number $[n/2]$ is the greatest integer $\leq n/2$ and the subscripted ωs are on $-1 < \omega < 1$.

As an example,

$$|H(j\omega)| = \frac{1}{\sqrt{1 + \epsilon^2 R_n^2(\omega)}} \tag{6.11}$$

for $n = 4$ might look like the response of Fig. 6.3 for some value of ϵ and the ω_k. The approximation satisfies the specifications

$$\begin{aligned} A_1 \leq |H(j\omega)| \leq 1, \qquad |\omega| \leq c \\ |H(j\omega)| \leq A_2, \qquad |\omega| \geq d \end{aligned} \tag{6.12}$$

with a transition band of $c \leq \omega \leq d$. (Note that by this definition, c will be in the passband if $A_1 > 1/\sqrt{2}$, in which case the pass- and transition bands overlap.)

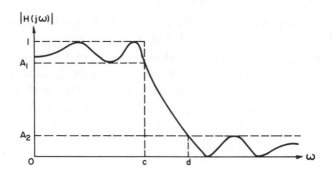

Figure 6.3. A rational function approximation to the ideal case.

To compare a more general filter with an all-pole filter, let us define the *order of complexity* of a filter as n, where $2n$ is the degree of the numerator or denominator, whichever is higher, of $f(\omega^2)$. Thus in the case of the Butterworth filter $(f(\omega^2) = \omega^{2n})$, the Chebyshev filter $(f(\omega^2) = \epsilon^2 C_n^2(\omega))$, the inverse Chebyshev filter $(f(\omega^2) = 1/\epsilon^2 C_n^2(1/\omega))$, and the filter defined by (6.8) and (6.9) or (6.10), the order of complexity is n.

For a given order of complexity and a given set of specifications A_1, A_2, and c, the filter is *optimal* if d (or the length $d - c$ of the transition band) is a minimum. As has been noted earlier, in the case of all-pole filters, the optimal filter is the Chebyshev filter. However, in the general case it is possible to improve on the Chebyshev filter in this respect. The optimal, as might be expected, occurs when the ω_k in (6.9) or (6.10) are chosen so that the amplitude function has equal passband ripples and equal stopband ripples [P-1]. In this case, the function $R_n(\omega)$ is called a *Chebyshev rational function*, and the ω_k are related to the *Jacobi elliptic sine functions*. The resulting filter is called the *elliptic filter*, and was first introduced by Cauer. We shall not undertake a discussion of the elliptic functions nor attempt to show that the elliptic filter is the optimal case. We shall consider instead a method of finding R_n without the use of elliptic functions and refer the reader to such works as Guillemin [G] and Weinberg [W].

As an example, the amplitude response of an elliptic filter is shown in Fig. 6.4 for the case $n = 5$, $A_1 = 0.9$, and $A_2 = 0.1$. As we shall see later, this requires that $\epsilon = 2.19521$ and yields a transition band of $0.940 < \omega < 1.064$.

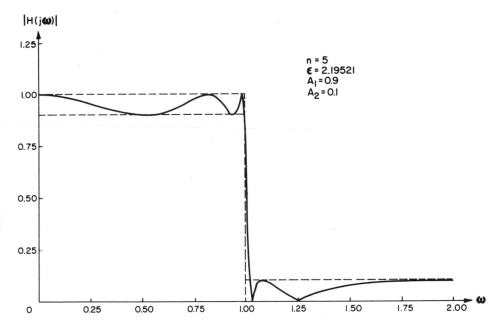

Figure 6.4. An elliptic filter amplitude response.

Since the elliptic filter is equiripple in both its passband and its stopband, referring to Fig. 6.3, we must have

$$A_1 = \frac{1}{\sqrt{1 + \epsilon^2 \delta^2}} \tag{6.13}$$

where δ is the maximum value of $|R_n(\omega)|$ on $0 \leq \omega \leq c$. Also we note from (6.9) and (6.10) that

$$R_n(1/\omega) = 1/R_n(\omega) \tag{6.14}$$

so that for $\omega > d$, the minimum value of $|R_n(\omega)|$ is $1/\delta$. Moreover, since at c we have $|R_n(\omega)| = \delta$ and at d we have $|R_n(\omega)| = 1/\delta$, it follows, in view of (6.14) that $d = 1/c$ and thus that $c < 1 < 1/c$ ($c = 1$ only in case $A_1 = A_2$, which is not acceptable). Therefore, since the stopband is also equiripple, we have

$$A_2 = \frac{1}{\sqrt{1 + \epsilon^2/\delta^2}} \tag{6.15}$$

The transition band is $c < \omega < 1/c$.

6.3 OBTAINING THE ELLIPTIC FILTER FUNCTIONS

In this section we outline a method [JJK] for obtaining the amplitude response of an elliptic filter of minimum complexity with a given pass-band ripple, a given stopband ripple, and a transition band within the interval $c < \omega < 1/c$ for some specified c. In other words, for given values of A_1, A_2, and c we shall find for minimum n a function with equal ripples on $0 < \omega < c_1 \geq c$ and on $\omega > 1/c_1 \leq 1/c$. The ripple widths on the two intervals will be respectively $1 - A_1$ and A_2. The details in the derivation of the method are given in reference [JJK] and will not be presented here.

We define the rational function $f_n(a, b; \omega)$ by

$$f_n(a, b; \omega) = \frac{P_n^{(a,a)}(b\omega)}{\omega^n P_n^{(a,a)}(b/\omega)} \tag{6.16}$$

where $P_n^{(a,a)}(x)$ is the *ultraspherical polynomial* of degree n with a parameter a [JJ]. The function f_n may also be written in the form

$$f_n(a, b; \omega) = \frac{\sum\limits_{i=0}^{[n/2]} B_{n-2i} \omega^{n-2i}}{\sum\limits_{i=0}^{[n/2]} B_{n-2i} \omega^{2i}} \tag{6.17}$$

where for $n = 2, 3, \ldots, 9$, the B_{n-2i} are given in Table 6.1.

From (6.13) and (6.15) we may solve for ϵ^2 and δ^2 obtaining

$$\epsilon^2 = \frac{\sqrt{(1 - A_1^2)(1 - A_2^2)}}{A_1 A_2}$$
$$\delta^2 = \frac{A_2\sqrt{1 - A_1^2}}{A_1\sqrt{1 - A_2^2}} \tag{6.18}$$

Thus for a given A_1 and A_2 we may find the required ϵ^2 and δ^2.

Using the value of δ found in (6.18) we select from Fig. 6.5 the minimum value of n which yields a value of c equal to or greater than the specified value. Then using these values of n and δ we obtain b from Fig. 6.6 and a from the appropriate one of Figs. 6.7, 6.8, and 6.9. Finally, using the values of n, a, and b we calculate $f_n(a, b; \omega)$ from (6.17) and Table 6.1. This function f_n is then $\pm R_n(\omega)$ to be used in (6.11) for the elliptic filter response $|H(j\omega)|$.

TABLE 6.1
Coefficients in $f_n(a, b; \omega)$

n	B_n	B_{n-2}	B_{n-4}	B_{n-6}	B_{n-8}
2	$2\left(a+\frac{3}{2}\right)b^2$	-1			
3	$2\left(a+\frac{5}{2}\right)b^3$	$-3b$			
4	$4\left(a+\frac{5}{2}\right)\left(a+\frac{7}{2}\right)b^4$	$-12\left(a+\frac{5}{2}\right)b^2$	3		
5	$4\left(a+\frac{7}{2}\right)\left(a+\frac{9}{2}\right)b^5$	$-20\left(a+\frac{7}{2}\right)b^3$	$15b$		
6	$8\left(a+\frac{7}{2}\right)\left(a+\frac{9}{2}\right)\left(a+\frac{11}{2}\right)b^6$	$-60\left(a+\frac{7}{2}\right)\left(a+\frac{9}{2}\right)b^4$	$90\left(a+\frac{7}{2}\right)b^2$	-15	
7	$8\left(a+\frac{9}{2}\right)\left(a+\frac{11}{2}\right)\left(a+\frac{13}{2}\right)b^7$	$-84\left(a+\frac{9}{2}\right)\left(a+\frac{11}{2}\right)b^5$	$210\left(a+\frac{9}{2}\right)b^3$	$-150b$	
8	$16\left(a+\frac{9}{2}\right)\left(a+\frac{11}{2}\right)\left(a+\frac{13}{2}\right)\left(a+\frac{15}{2}\right)b^8$	$-224\left(a+\frac{9}{2}\right)\left(a+\frac{11}{2}\right)\left(a+\frac{13}{2}\right)b^6$	$840\left(a+\frac{9}{2}\right)\left(a+\frac{11}{2}\right)b^4$	$-840\left(a+\frac{9}{2}\right)b^2$	105
9	$16\left(a+\frac{11}{2}\right)\left(a+\frac{13}{2}\right)\left(a+\frac{15}{2}\right)\left(a+\frac{17}{2}\right)b^9$	$-288\left(a+\frac{11}{2}\right)\left(a+\frac{13}{2}\right)\left(a+\frac{15}{2}\right)b^7$	$1512\left(a+\frac{11}{2}\right)\left(a+\frac{13}{2}\right)b^5$	$-2520\left(a+\frac{11}{2}\right)b^3$	$945b$

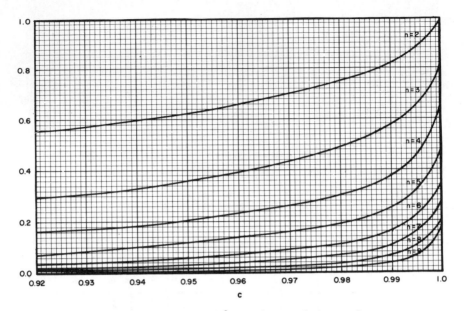

Figure 6.5. Curves of δ vs. c for $n = 2, 3, \ldots, 9$.
(*Courtesy IEEE*)

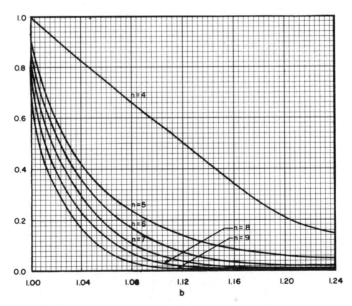

Figure 6.6. Curves of δ vs. b for $n = 4, 5, \ldots, 9$.
(*Courtesy IEEE*)

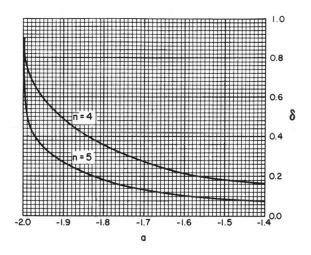

Figure 6.7. Curves of δ vs. a for $n = 4$ and 5.

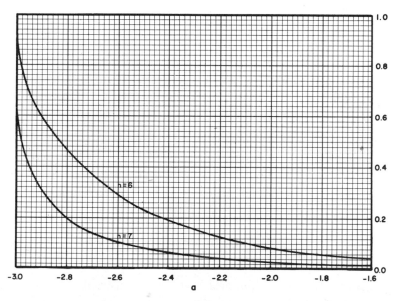

Figure 6.8. Curves of δ vs. a for $n = 6$ and 7.
(*Courtesy IEEE*)

125

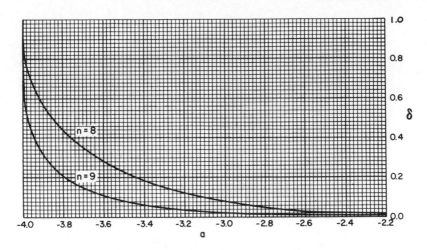

Figure 6.9. Curves of δ vs. a for $n = 8$ and 9.

The cases $n = 2$ and 3 are not shown in the graphs because a and b are not unique for these cases. They are considered later in Exercises 6.6 and 6.7, where it will be helpful to note that in the general case, their limits are given by

$$-[n/2] < a < -1/2, \qquad b > 1 \qquad (6.19)$$

As an example, suppose we want an elliptic filter of least complexity having $A_1 = 0.9$, $A_2 = 0.1$, and $c \geq 0.99$. By (6.18) we may obtain

$$\epsilon^2 = 4.819, \qquad \delta = 0.221$$

From Fig. 6.5, for the calculated value of δ and the required value of c, we see that the minimum order is $n = 6$ (corresponding to $c = 0.9945$). From Figs. 6.6 and 6.8 we have

$$b = 1.066, \qquad a = -2.47$$

Using these values of ϵ^2, n, a, and b we may construct $|H(j\omega)|$ from (6.11) and Table 6.1. A plot of the amplitude function obtained is shown in Fig. 6.10.

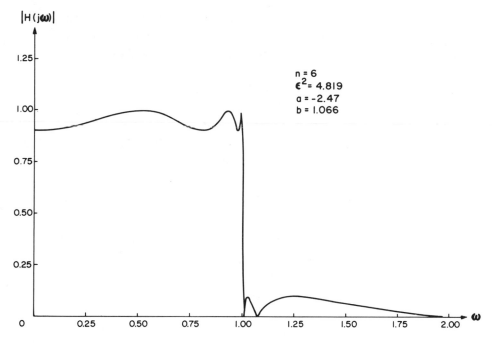

Figure 6.10. An elliptic amplitude response with $A_1 = 0.9$ and $A_2 = 0.1$.

6.4   FORM OF THE ELLIPTIC FILTER TRANSFER FUNCTION

The general form of the elliptic filter transfer function $H(s)$ may be obtained from

$$H(s)H(-s) = \frac{1}{1 + \epsilon^2 R_n^2(\omega)}\Bigg|_{\omega^2 = -s^2} \qquad (6.20)$$

In the case n even, $[n/2] = n/2$ and by (6.9) and (6.20) we have

$$H(s)H(-s) = \frac{1}{1 + \epsilon^2 \left[\displaystyle\prod_{i=1}^{n/2} \frac{(\omega_{2i-1}^2 + s^2)}{(1 + \omega_{2i-1}^2 s^2)} \right]^2}$$

$$= \frac{\displaystyle\prod_{i=1}^{n/2} (1 + \omega_{2i-1}^2 s^2)^2}{\displaystyle\prod_{i=1}^{n/2} (1 + \omega_{2i-1}^2 s^2)^2 + \epsilon^2 \displaystyle\prod_{i=1}^{n/2} (\omega_{2i-1}^2 + s^2)^2} \qquad (6.21)$$

The denominator of (6.21) is of degree $2n$ and thus the denominator of $H(s)$ will be of degree n, and will of course be Hurwitz. The numerator of $H(s)$ will be $\prod_{i=1}^{n/2}(1 + \omega_{2i-1}^2 s^2)$. Thus for n even we have

$$H(s) = \frac{K \prod_{i=1}^{n/2} (s^2 + a_{2i-1}^2)}{s^n + b_{n-1}s^{n-1} + \cdots + b_1 s + b_0} \tag{6.22}$$

where the b_i are determined by selecting the Hurwitz factor in the denominator of $H(s)H(-s)$ given in (6.21), and

$$a_{2i-1} = \frac{1}{\omega_{2i-1}} \tag{6.23}$$

For n odd, $[n/2] = (n-1)/2$ and using (6.10) we may show that

$$H(s) = \frac{K \prod_{i=1}^{(n-1)/2} (s^2 + a_{2i}^2)}{s^n + b_{n-1}s^{n-1} + \cdots + b_1 s + b_0} \tag{6.24}$$

where

$$a_{2i} = \frac{1}{\omega_{2i}} \tag{6.25}$$

The b_i are obtained as before and, of course, are different for n odd. We may note that for n even $H(s)$ is a ratio of nth degree polynomials, whereas for n odd, it is an $(n-1)$th degree polynomial divided by an nth degree polynomial. In both cases K is a constant resulting from making the denominator polynomial monic. In either the odd or even case, the synthesis of the elliptic filter can be carried out passively using the zero-shifting techniques described earlier in Sec. 5.6.

As examples, if $n = 2$ we have

$$H(s) = \frac{K(s^2 + a_1^2)}{s^2 + b_1 s + b_0} \tag{6.26}$$

and for $n = 3$,

$$H(s) = \frac{K(s^2 + a_2^2)}{s^3 + b_2 s^2 + b_1 s + b_0} \tag{6.27}$$

In a simple case we may find the transfer function without factoring $H(s)H(-s)$ by noting that $|H(j0)| = H(0)$ and comparing terms in $|H(j\omega)|$ obtained from $H(s)$ and from (6.11) for a given ϵ^2 and ω_k^2. The reader is asked, for example, to show in Exercise 6.8 that for $n = 2$, we have in (6.26)

$$a_1 = \frac{1}{\omega_1^2}$$

$$b_0^2 = \frac{1 + \epsilon^2 \omega_1^4}{\epsilon^2 + \omega_1^4} \tag{6.28}$$

$$b_1^2 = 2\left[\sqrt{\frac{1 + \epsilon^2 \omega_1^4}{\epsilon^2 + \omega_1^4}} - \frac{\omega_1^2(1 + \epsilon^2)}{\epsilon^2 + \omega_1^2}\right]$$

6.5 SUMMARY

Transfer functions with finite zeros were discussed in this chapter, and two examples, the inverse Chebyshev and elliptic filters, were considered in detail. The former has a maximally-flat passband response while the latter has ripples in both its pass- and stopbands. The elliptic filter is of great importance because it is the optimal approximation to the ideal low-pass response.

A method was presented for obtaining elliptic filter amplitude responses that did not require a knowledge of elliptic function theory. The method yields the amplitude function of least complexity, having a specified passband ripple, stopband ripple, and transition band.

In the next chapter we shall consider another network function having finite zeros—namely the all-pass filter. Also we shall complete our study of approximation theory by developing constant-time-delay filter functions.

EXERCISES

6.1. Show that the cutoff point ω_c for the low-pass inverse Chebyshev filter is given by

$$\omega_c = \frac{1}{\cosh\left(\frac{1}{n}\cosh^{-1}\frac{1}{\epsilon}\right)}$$

and hence if $0 < \epsilon < 1$, then $0 < \omega_c < 1$.

6.2. (a) Show that if $|H(j\omega)|^2$ shown in Fig. Ex. 6.2 (a) below is the amplitude-squared function of an inverse Chebyshev filter, then its corresponding Chebyshev filter has the amplitude-squared function

$|H_c(j\omega)|^2$ shown in Fig. Ex. 6.2 (b). Thus to fit an inverse Chebyshev filter to the specifications of (a) we need only fit a Chebyshev filter to the specifications of (b), which may be used to determine n and ϵ^2.

(b)   Use the results of part (a) to obtain $|H(j\omega)|^2$ for the inverse Chebyshev filter with minimum n which satisfies the requirements of Exercise 3.3. Compare with the results obtained there for the Butterworth and Chebyshev filters.

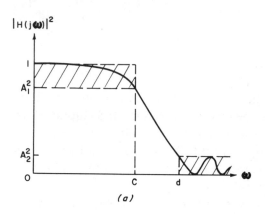

(a)

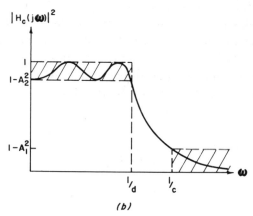

(b)

Figure Ex. 6.2.

6.3.   The functions

(a) $R_2(\omega) = \dfrac{0.849^2 - \omega^2}{1 - 0.849^2\omega^2}$

and

(b) $R_3(\omega) = \dfrac{\omega(0.930^2 - \omega^2)}{1 - 0.930^2\omega^2}$

are Chebyshev rational functions for which $c = \sqrt[4]{0.9}$. Verify this by sketching these functions and finding δ.

6.4. Find a, b, ϵ^2, and n for the elliptic filter of least complexity for which $A_1 = 0.9$, $A_2 = 0.05$, and c is at least 0.98.

$$\text{Ans. } n = 6, \ c = 0.988, \ a = -2.3,$$
$$b = 1.084.$$

6.5. Find the minimum order n of a Chebyshev filter which fits the specifications of Exercise 6.4. (Note that for minimum n the transition band is of maximum length. Thus take $c = 0.98$ and scale to $c = 1$. This yields $|H(j\omega)| \le A_2$ for $\omega \ge (1/0.98)^2 = 1.0412$, and does not affect n.)

$$\text{Ans. } n = 16.$$

6.6. The parameters a and b satisfy, for $n = 2$ and 3 respectively [JJK],

$$(2a + 3)b^2 = \frac{1}{\alpha^2}$$

and

$$(2a + 5)b^2 = \frac{3}{\alpha^2}$$

where

$$R_2(\omega) = \frac{\omega^2 - \alpha^2}{1 - \alpha^2\omega^2}$$

and

$$R_3(\omega) = \frac{\omega(\omega^2 - \alpha^2)}{1 - \alpha^2\omega^2}$$

and $b > 1$, $-1 < a < -1/2$. For $n = 2$, sketch $R_2(\omega)$ and show that

$$\alpha^2 = \delta, \qquad c = \sqrt{2\delta/(1 + \delta^2)}$$

For $A_1 = 0.9$, $A_2 = 0.1$, and $c \ge 0.62$, show that $n = 2$ is sufficient, and in that case we may take $a = -3/4$ and $b^2 = 3.0133$.

6.7. Using the results given in Exercise 6.6 for $n = 3$, sketch $R_3(\omega)$ and show that

$$\alpha^2 = \frac{c^3 - \delta}{c(1 - \delta c)}$$

For $A_1 = 0.8$, $A_2 = 0.2$, and $c \ge 0.95$, show that the minimum order is $n = 3$ and that we may take $a = -0.9$ and $b = 1.0671$.

6.8. Verify (6.28).

6.9. Obtain a passive realization consisting of a singly-terminated LC network, of the inverse Chebyshev function of Sec. 6.1,

$$\frac{V_2}{V_1} = \frac{K(3s^2 + 4)}{s^3 + 2.5s^2 + 2s + 2}$$

6.10. Show that for the case $n = 2$ in Exercise 6.6, the Chebyshev rational function is given by

$$R_2(\omega) = \frac{0.221 - \omega^2}{1 - 0.221\omega^2}$$

and the transfer function is given by

$$H(s) = \frac{K(s^2 + 4.525)}{s^2 + 0.692s + 0.504}$$

7

Phase-Shifting and Time-Delay Filters

7.1 INTRODUCTION

Up to now the filters we have been interested in are frequency-selective filters, in which the amplitude response is the important factor. In this chapter we shall consider filters in which the phase response and/or its associated time delay are the factors of interest.

The filters we shall consider are all-pass filters, for which the amplitude response is constant and the phase response is a function of frequency, and Bessel filters, which have a maximally-flat time delay. Finally, in Sec. 7.8 we shall consider a filter that has both the all-pass and maximally-flat time delay properties.

The all-pass filter has a nonminimum phase, which has been defined in Sec. 2.5 as a filter whose transfer function has right-half-plane zeros. Before proceeding to the all-pass filter, we shall consider first some remarks about amplitude and phase, particularly nonminimum phase functions.

7.2 AMPLITUDE AND PHASE FUNCTIONS

As we have seen in Sec. 2.1, if a network function represents a stable network then it can have no right-half-plane poles. If the function is a driving-point function, then its reciprocal is also a network function, indicating that neither its poles nor zeros may be in the right-half plane. However, since the reciprocal of a transfer function is not in general a network function, right-half plane zeros, and hence nonminimum phase functions, are permissible.

To see why the terms "minimum" and "nonminimum" are applied to the phase, let us consider a graphical method of evaluating the magnitude and phase of a transfer function. Suppose $H(s)$ in factored form is given by

$$H(s) = \frac{K(s - z_1)(s - z_2) \cdots (s - z_m)}{(s - p_1)(s - p_2) \cdots (s - p_n)} = \frac{KP(s)}{Q(s)} \qquad (7.1)$$

where z_i, $i = 1, 2, \ldots, m$, and p_j, $j = 1, 2, \ldots, n$, are respectively the finite zeros and poles. If $s = j\omega$ we may write $H(j\omega)$ in the form

$$H(j\omega) = \frac{KM_1 M_2 \cdots M_m}{N_1 N_2 \cdots N_n} e^{j(\alpha_1 + \alpha_2 + \cdots + \alpha_m - \beta_1 - \beta_2 - \cdots - \beta_n)} \qquad (7.2)$$

where

$$j\omega - z_i = M_i e^{j\alpha_i}, \qquad i = 1, 2, \ldots, m$$
$$j\omega - p_i = N e^{j\beta_i}, \qquad i = 1, 2, \ldots, n \qquad (7.3)$$

In general, $H(j\omega) = |H(j\omega)| e^{j\phi(\omega)}$, and thus we may compare (7.2) and (7.3) to obtain the amplitude function

$$|H(j\omega)| = \frac{KM_1 M_2 \cdots M_m}{N_1 N_2 \cdots N_n} \qquad (7.4)$$

and the phase function

$$\phi(\omega) = \alpha_1 + \alpha_2 + \cdots + \alpha_m - (\beta_1 + \beta_2 + \cdots + \beta_n) \qquad (7.5)$$

We may interpret the term $j\omega - z_i$ as a vector in the s-plane from z_i to $j\omega$, as shown in Fig. 7.1. The length or amplitude of the vector is $|j\omega - z_i| = M_i$, and its direction or phase is α_i. A similar interpretation may be given for $j\omega - p_i$, shown also in Fig. 7.1. Thus $|H(j\omega)|$ and $\phi(\omega)$ could be measured graphically by plotting all the vectors corresponding to numerator and denominator terms in $H(j\omega)$ and making the calculations indicated in (7.4) and (7.5).

Fig. 7.2 shows a left-half plane zero $z_1 = -\sigma_1 + j\omega_1$ and its *mirror*

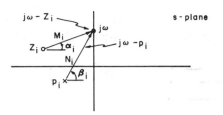

Figure 7.1. Geometrical interpretation of Eq. (7.3).

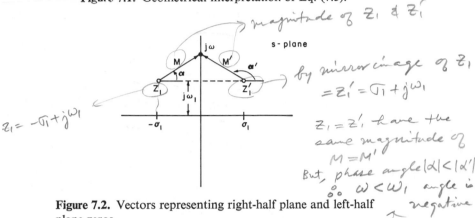

[handwritten: magnitude of Z_1 & Z_1']

[handwritten: by mirror image of Z_1 = $Z_1' = \sigma_1 + j\omega_1$]

[handwritten: $Z_1 = Z_1'$ have the same magnitude of $M = M'$. But, phase angle $|\alpha| < |\alpha'|$. $\therefore \omega < \omega_1$, angle is negative]

[handwritten: $Z_1 = -\sigma_1 + j\omega_1$]

Figure 7.2. Vectors representing right-half plane and left-half plane zeros.

image, a right-half plane zero $z_1' = \sigma_1 + j\omega_1$. Evidently the vectors $j\omega - z_1$ and $j\omega - z_1'$ shown, have the same magnitudes, $M = M'$, but in the case of the phase angles, we have $|\alpha| < |\alpha'|$. These same results hold if $\omega < \omega_1$, in which case the angles are negative. Thus the motivations are clear for the definitions "minimum" and "nonminimum" phase applied respectively to functions without and with right-half plane zeros.

Suppose now we have a transfer function $H(s)$ with all left-half plane poles and whose zeros are mirror images of the poles, and thus are all in the right-half plane. Each pole and its mirror image zero constitute a pair like z_1 and z_1' in Fig. 7.2, and therefore their amplitudes cancel, as may be seen in (7.4). Since also $m = n$, we have by (7.4),

$$|H(j\omega)| = K \qquad (7.6)$$

[handwritten: by canceling magnitude of M & H', $\frac{M}{M=M'}$ by eq. 7.4 $M=N$]

and thus the function is that of an all-pass filter. Since the pole-zero pairs are either negatives of each other, in case they are real, or consti-

[handwritten: $\therefore$ all pass filter]

tute half of a quad if they are complex, we must have in (7.1), $P(s) = Q(-s)$. Therefore, in general an all-pass filter function has the form

FORM OF
ALL-PASS FILTER

$$H(s) = K\frac{Q(-s)}{Q(s)} \tag{7.7}$$

As examples, consider the functions

$$H(s) = K\frac{s - a}{s + a} \tag{7.8}$$

and

$$H(s) = K\frac{s^2 - as + b}{s^2 + as + b} \tag{7.9}$$

The pole-zero plots are shown in Fig. 7.3, where it may be seen that in (7.8) we have

$$|H(j\omega)| = K\frac{M_1}{M_1} = K$$

and in (7.9) we have

$$|H(j\omega)| = K\frac{M_1 M_2}{M_1 M_2} = K$$

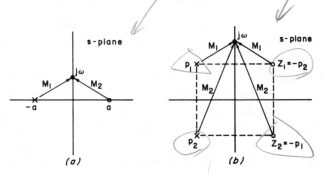

Figure 7.3. Pole-zero plots of (a) Eq. (7.8), and (b) Eq. (7.9).

7.3 ALL-PASS FILTERS

As was pointed out in the previous section, the transfer function of an all-pass filter is given in general by

$$H(s) = K\frac{Q(-s)}{Q(s)} \tag{7.10}$$

where K is a constant and $Q(s)$ is a Hurwitz polynomial. Equation (7.10) is not an approximation since $|H(j\omega)| = K$, which is an exact all-pass amplitude function. In the passive three-terminal case we must have $|K| \leq 1$ (see Exercise 5.9), and as we shall see in Sec. 7.5, if $K = 1$, the passive realization can be carried out readily by means of symmetrical LC lattices with a 1Ω resistor termination.

The primary purpose of an all-pass filter is for phase shift. If we have a filter whose amplitude response is adequate but whose phase response needs correction, it is possible to cascade the filter with an all-pass section, retaining the desired amplitude response, but adding to the phase response the phase shift of the all-pass filter. That is, if

$$H(j\omega) = |H(j\omega)|e^{j\phi(\omega)}$$

is the transfer function of the filter to be cascaded with an all-pass filter having a transfer function

$$H_A(j\omega) = 1e^{j\phi_A(\omega)}$$

then the overall function is

$$H(j\omega)H_A(j\omega) = |H(j\omega)|e^{j[\phi(\omega) + \phi_A(\omega)]}$$

We are assuming, of course, that the cascading process does not change either of the transfer functions.

The phase response of the all-pass filter defined by (7.10) is given by

$$\phi(\omega) = -2\phi_Q(\omega)$$

where $\phi_Q(\omega)$ is the phase response of the polynomial

$$Q(s) = M(s) + N(s) \tag{7.11}$$

Since M is even and N is odd, we have

$$\phi(\omega) = -2\tan^{-1}\left[\frac{N(j\omega)}{jM(j\omega)}\right] \tag{7.12}$$

For example, a second-order all-pass response is given by

$$H_n(s) = \frac{s^2 - as + b}{s^2 + as + b} \tag{7.13}$$

for which the phase response is

$$\phi_n(\omega) = -2\tan^{-1}\left[\frac{a\omega}{b-\omega^2}\right] \tag{7.14}$$

The impulse response, $h_n(t) = \mathcal{L}^{-1}H_n(s)$, is given by

$$h_n(t) = \delta(t) - 2ae^{-\zeta_n t}\left(\cos kt - \frac{a}{2k}\sin kt\right)$$

where $\delta(t)$ is the unit impulse function, k is given by

$$k = \sqrt{b - a^2/4} \tag{7.15}$$

and

$$\zeta_n = a/2 \tag{7.16}$$

is the *damping constant.*

If $H_n(s)$ in (7.13) is a normalized response, where a and b are numbers like 1 or 2, etc., then we may frequency scale its network, replacing s by s/ω_0, and obtain the transfer function

$$H(s) = H_n(s/\omega_0) = \frac{s^2 - a\omega_0 s + b\omega_0^2}{s^2 + a\omega_0 s + b\omega_0^2} \tag{7.17}$$

In this case the denormalized functions, $\phi(\omega) = \phi_n(\omega/\omega_0)$ and $h(t) = \mathcal{L}^{-1}H(s)$, are given by

$$\phi(\omega) = -2\tan^{-1}\left[\frac{a\omega_0\omega}{b\omega_0^2 - \omega^2}\right] \tag{7.18}$$

and

$$h(t) = \omega_0\left[\delta(t) - 2ae^{-\zeta t}\left(\cos k\omega_0 t - \frac{a}{2k}\sin k\omega_0 t\right)\right] \tag{7.19}$$

where k is given by (7.15) and the damping constant is

$$\zeta = \frac{a\omega_0}{2} \tag{7.20}$$

Thus if we desire a certain phase shift ϕ_0 at a certain frequency ω_0, and a damping constant ζ, then by (7.20) we may find a and by (7.18) we have

$$\phi_0 = -2\tan^{-1}\left[\frac{a}{b-1}\right] \tag{7.21}$$

which determines b. (We note that (7.21) is also (7.14) for $\omega = 1$.)

Alternately, we may design a normalized all-pass filter to have a certain phase shift ϕ_0 and damping constant ζ_n at $\omega = 1$ rad/s. By (7.16) and (7.21) we determine a and b and thus the transfer function $H_n(s)$ given by (7.13). Frequency scaling with a scale factor ω_0 then results in the denormalized filter with the given ϕ_0 and $\zeta = \zeta_n\omega_0$ at $\omega = \omega_0$.

Higher-order all-pass responses may be used for more flexibility in obtaining phase shift and damping constants.

7.4   LATTICE REALIZATIONS

A nonminimum phase function cannot be realized by a passive ladder network because the series and shunt elements cannot become open and short circuits at right-half plane frequencies. However, if the numerator of the function is odd or even, the realization may be carried out by means of a symmetrical lattice, the general notation for which is shown in Fig. 7.4. As discussed in Exercise 2.9, the network of Fig. 7.4 is a representation of a lattice having *series* arms of impedance Z_a between terminals 1 and 2 and between 1' and 2', and having *cross* arms of impedance Z_b between terminals 1 and 2' and between 1' and 2.

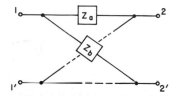

Figure 7.4. A symmetrical lattice.

Also, as discussed in Exercise 2.9, the z-parameters of the symmetrical lattice are given by

$$z_{11} = z_{22} = \tfrac{1}{2}(Z_b + Z_a)$$
$$z_{12} = z_{21} = \tfrac{1}{2}(Z_b - Z_a) \tag{7.22}$$

The y-parameters may be found from the z-parameters or from the lattice directly, and may be shown to be

$$y_{11} = y_{22} = \tfrac{1}{2}(Y_b + Y_a)$$
$$y_{12} = y_{21} = \tfrac{1}{2}(Y_b - Y_a) \tag{7.23}$$

where $Y_a = 1/Z_a$ and $Y_b = 1/Z_b$. From (7.22) and (7.23) we may solve for $Z_a, Z_b, Y_a,$ and Y_b, obtaining

$$Z_a = z_{11} - z_{12} = z_{22} - z_{21}$$
$$Z_b = z_{11} + z_{12} = z_{22} + z_{21}$$

(7.24)

and

$$Y_a = y_{11} - y_{12} = y_{22} - y_{21}$$
$$Y_b = y_{11} + y_{12} = y_{22} + y_{21}$$

(7.25)

The presence of the minus signs in (7.24) and (7.25) enables us to realize right-half plane zeros with the symmetrical lattice. As an example, let us consider the function

$$\frac{V_2}{V_1} = \frac{Ks(s^2 - 4)}{s^3 + 2s^2 + 2s + 1}$$

(7.26)

or

$$\frac{V_2}{V_1} = \frac{\left[\dfrac{Ks(s^2 - 4)}{2s^2 + 1}\right]}{1 + \dfrac{s^3 + 2s}{2s^2 + 1}}$$

from which we identify

$$y_{22} = \frac{s(s^2 + 2)}{2s^2 + 1}$$

$$-y_{21} = \frac{Ks(s^2 - 4)}{2s^2 + 1}$$

(7.27)

We note that the function V_2/V_1, as well as $-y_{21}$, has zeros at $s = 0$, ± 2.

Expanding (7.27) in partial fractions yields

$$y_{22} = \frac{s}{2} + \frac{\frac{3}{2}s}{2s^2 + 1}, \qquad -y_{21} = \frac{Ks}{2} - \frac{\frac{9K}{2}s}{2s^2 + 1}$$

and by (7.25) we have

$$Y_a = \frac{(1 + K)s}{2} + \frac{3(1 - 3K)s}{2(2s^2 + 1)}$$

$$Y_b = \frac{(1 - K)s}{2} + \frac{3(1 + 3K)s}{2(2s^2 + 1)}$$

Both Y_a and Y_b are _LC_ admittances if we have $0 < 3K \le 1$. Selecting $K = 1/3$ eliminates two elements in Y_a and results in

sub. K = ⅓
& solve eqn.

$$Y_a = 2s/3, \quad Y_b = s/3 + \cfrac{1}{\cfrac{2}{3}s + \cfrac{1}{3s}}$$

The network is shown in Fig. 7.5 with the required 1Ω termination. Analysis shows that (7.26) is realized with $K = 1/3$.

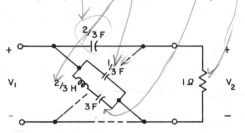

Figure 7.5. A lattice realization of Eq. (7.26).

In the general case y_{22} and $-y_{21}$ may be expanded in partial fractions and Y_a and Y_b formed in accordance with (7.25). If $H = V_2/I_1$, a dual development is carried out using z_{22} and z_{21}. Then the constant K is selected to make the function realizable. If the partial fraction expansions are used to obtain the lattice elements, the limits on K are easier to see.

In some cases the transfer function

$$H(s) = \frac{P(s)}{M_2(s) + N_2(s)}$$

where $P(s)$ is neither even nor odd, may be modified so as to be realizable by a symmetrical lattice. For example, if $P(s) = M_1(s) - N_1(s)$, where $P(-s) = M_1(s) + N_1(s)$ is Hurwitz, then we may write $H(s)$ in the form

$$H(s) = \frac{M_1(s) - N_1(s)}{M_2(s) + N_2(s)} \cdot \frac{M_1(s) + N_1(s)}{M_1(s) + N_1(s)}$$

$$= \frac{M_1^2(s) - N_1^2(s)}{Q(s)}$$

where $Q(s) = (M_1 + N_1)(M_2 + N_2)$ is Hurwitz. Since the numerator is now even, the previous methods apply.

Finally suppose we have

$$H(s) = \frac{P(s)}{Q(s)} = \frac{P_1(s)P_2(s)}{Q(s)}$$

where P_1 is even or odd, P_2 is neither even nor odd but is Hurwitz, and $|H(j\omega)|$ is the desired amplitude. Then we may write

$$H'(s) = \frac{P_1(s)P_2(s)}{Q(s)} \cdot \frac{P_2(-s)}{P_2(s)}$$

where now the numerator is even or odd, as desired, and $Q(s)P_2(s)$ is the denominator, which is Hurwitz. The foregoing methods apply to $H'(s)$ and since

$$|H'(j\omega)| = H(j\omega)| \cdot \left|\frac{P_2(-j\omega)}{P_2(j\omega)}\right| = |H(j\omega)|$$

it has the desired amplitude.

7.5 CONSTANT-RESISTANCE LATTICES

A two-port network is a *constant-resistance* network if when the 2–2' port is terminated in a resistance R, the input impedance seen at the 1–1' port is also R, as illustrated in Fig. 7.6. In the case $R = 1\Omega$ we

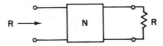

Figure 7.6. A constant-resistance network N.

have, using the result of Exercise 2.8(b),

$$\frac{z_{11} + \Delta_z}{z_{22} + 1} = 1 \tag{7.28}$$

where $\Delta_z = z_{11}z_{22} - z_{12}z_{21}$. In the case of the symmetrical lattice we have

$$\Delta_z = [\tfrac{1}{2}(Z_b + Z_a)]^2 - [\tfrac{1}{2}(Z_b - Z_a)]^2 = Z_a Z_b \tag{7.29}$$

so that (7.28) may be written, since $z_{11} = z_{22}$,

$$Z_a Z_b = 1 \qquad (7.30)$$

Defining $G_{21}(s) = V_2(s)/V_1(s)$, we have for the lattice terminated in 1Ω,

$$G_{21} = \frac{-y_{21}}{1 + y_{22}} = \frac{-\frac{1}{2}(Y_b - Y_a)}{1 + \frac{1}{2}(Y_b + Y_a)} = \frac{Z_b - Z_a}{2Z_a Z_b + Z_a + Z_b}$$

Eliminating Z_b by (7.30) results in

$$G_{21} = \frac{1 - Z_a^2}{2Z_a + Z_a^2 + 1} = \frac{1 - Z_a^2}{(1 + Z_a)^2}$$

or

$$G_{21}(s) = \frac{V_2(s)}{V_1(s)} = \frac{1 - Z_a(s)}{1 + Z_a(s)} \qquad (7.31)$$

where

$$Z_b(s) = \frac{1}{Z_a(s)} \qquad (7.32)$$

Equation (7.31) may also be solved for Z_a, resulting in

$$Z_a = \frac{1 - G_{21}}{1 + G_{21}} \qquad (7.33)$$

Thus given G_{21}, we may find Z_a directly by (7.33) and Z_b by (7.32). It may, of course, be necessary to multiply G_{21} by a suitable scale factor to make Z_a a realizable driving-point function.

Since the foregoing applies to a constant-resistance lattice with input impedance of 1Ω, we have $V_1 = I_1$, and therefore

$$\frac{V_2}{I_1} = \frac{V_2}{V_1} = G_{21}$$

Thus (7.32) and (7.33) apply also to the transfer function V_2/I_1.

Now let us consider a cascading of n two-port constant-resistance networks with $R = 1\Omega$, terminated in 1Ω, as shown in Fig. 7.7. Each two-port is loaded with 1Ω, either in the form of a network with 1Ω input resistance or a 1Ω load, as in the case of network N_n. Therefore the individual transfer functions are not affected by the loading if they

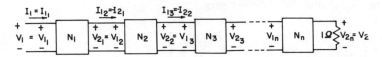

Figure 7.7. A cascading of two-port, constant-resistance networks.

are all derived for 1Ω terminations. We may write V_2/V_1 in the form

$$\frac{V_2}{V_1} = \frac{V_{2_n}}{V_{1_1}} = \frac{V_{2_1}}{V_{1_1}} \frac{V_{2_2}}{V_{1_2}} \frac{V_{2_3}}{V_{1_3}} \cdots \frac{V_{2_n}}{V_{1_n}}$$

since $V_{2_1} = V_{1_2}$, $V_{2_2} = V_{1_3}$, etc. Thus we have

$$\frac{V_2}{V_1} = G_{21}^{(1)} G_{21}^{(2)} \cdots G_{21}^{(n)} \tag{7.34}$$

where $G_{21}^{(i)} = V_{2_i}/V_{1_i}$, $i = 1, 2, \ldots, n$.

As the foregoing indicates, it may be possible to factor a given G_{21} into n simpler factors, in which case the problem of realizing a complicated two-port network can be solved by realizing n simpler two-ports by means of (7.32) and (7.33). Cascading the n two-ports as shown in Fig. 7.7 with a 1Ω termination then solves the problem.

In the case of all-pass filters, the use of constant-resistance lattices leads in general to LC networks terminated in a resistor. To see this, consider the general all-pass function,

$$\frac{V_2(s)}{V_1(s)} = \frac{Q(-s)}{Q(s)} = \frac{M(s) - N(s)}{M(s) + N(s)} \tag{7.35}$$

where $Q(s)$ is Hurwitz. (We are taking the case $K = 1$.) The even and odd parts of $Q(s)$ are respectively $M(s)$ and $N(s)$, so that $Q(-s)$ is as indicated. Writing (7.35) in the form

$$\frac{V_2}{V_1} = \frac{1 - \dfrac{N(s)}{M(s)}}{1 + \dfrac{N(s)}{M(s)}}$$

and comparing the result with (7.31) we see that $Z_a = N/M$, which is guaranteed to be LC, as is $Z_b = 1/Z_a = M/N$.

As an example, let us synthesize the function

$$\frac{V_2}{V_1} = \frac{(s-1)(s^2 - 2s + 2)}{(s+1)(s^2 + 2s + 2)} \qquad (7.36)$$

which is evidently an all-pass function. We may write the function as

$$\frac{V_2}{V_1} = G_{21}^{(1)} G_{21}^{(2)}$$

where

$$G_{21}^{(1)} = \frac{s-1}{s+1} = \frac{1 - 1/s}{1 + 1/s}$$

and

$$G_{21}^{(2)} = \frac{s^2 - 2s + 2}{s^2 + 2s + 2} = \frac{1 - \dfrac{2s}{s^2 + 2}}{1 + \dfrac{2s}{s^2 + 2}}$$

Thus for network 1 we have

$$Z_a^{(1)} = 1/Z_b^{(1)} = 1/s$$

and for network 2 we have

$$Z_a^{(2)} = 1/Z_b^{(2)} = \frac{2s}{s^2 + 2}$$

The networks cascaded and terminated in 1Ω to yield (7.36) are shown in Fig. 7.8.

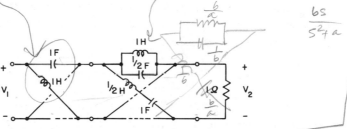

Figure 7.8. A realization of Eq. (7.36).

7.6 TIME DELAY

In the previous sections we have considered filters for which the characteristic of principal interest is the amplitude response, or in the case of the all-pass filter, is the phase response. In the next section we shall consider a filter for which we are primarily interested in the time delay, defined earlier in Sec. 1.2 by

$$T(\omega) = -\frac{d}{d\omega}\phi(\omega) \text{ sec} \qquad (7.37)$$

To see why time delay is important, let us consider a network with input $x(t)$ and output $y(t) = Kx(t - \tau)$, where K and τ are constants. The output is thus the input shifted in time and possibly amplified. The output signal, in this case, is defined to be *undistorted*. In the frequency domain we have

$$Y(s) = \mathscr{L}Kx(t - \tau) = Ke^{-s\tau}X(s)$$

and thus the transfer function is given by

$$H(s) = \frac{Y(s)}{X(s)} = Ke^{-s\tau} \qquad (7.38)$$

so that

$$|H(j\omega)| = K, \qquad \phi(\omega) = -\omega\tau$$

Therefore, in the distortionless case the time delay, defined by (7.37), is $T(\omega) = \tau$, a constant.

In the next section we shall obtain a realizable approximation to a constant time delay filter function, given by (7.38). To facilitate matters, and since the value of K is unimportant as far as $T(\omega)$ is concerned, we shall seek an approximation $H_n(s)$ to the normalized ($\tau = 1$) ideal response e^{-s}. That is,

$$H_n(s) \approx e^{-s} \qquad (7.39)$$

The phase response in the nonideal normalized case will be denoted by $\phi_n(\omega)$, and the time delay by $T_n(\omega)$, which will be given by

$$T_n(\omega) = -\frac{d}{d\omega}\phi_n(\omega) \qquad (7.40)$$

If we denormalize the function $H_n(s)$ by using a frequency scale factor ω_0, we have

$$H(s) = H_n(s/\omega_0), \quad \phi(\omega) = \phi_n(\omega/\omega_0) \qquad (7.41)$$

and

$$T_n(\omega/\omega_0) = -\frac{d}{d(\omega/\omega_0)}\phi_n(\omega/\omega_0)$$

$$= \omega_0\left[-\frac{d}{d\omega}\phi(\omega)\right]$$

$$= \omega_0 T(\omega)$$

Therefore we have

$$T(\omega) = \frac{1}{\omega_0} T_n(\omega/\omega_0) \tag{7.42}$$

or

$$T(\omega_0) = \frac{1}{\omega_0} T_n(1) \text{ sec} \tag{7.43}$$

We see from the foregoing discussion that if we could obtain a normalized network with a time delay approximating $\tau = 1s$ over the interval $0 \leq \omega \leq 1$, then frequency scaling the network by a factor ω_0 yields an approximately constant time delay of $\tau = 1/\omega_0 s$ over the interval $0 \leq \omega \leq \omega_0$.

7.7 BESSEL FILTERS

Let us now turn our attention to obtaining a realizable approximation to the ideal constant time delay function e^{-s}. The method we outline was given by Storch [St] and the network which is obtained is called a *Bessel* filter, for reasons which will be noted later.

We begin by noting the identities

$$e^{-s} = \frac{1}{e^s} = \frac{1}{\cosh s + \sinh s} \tag{7.44}$$

The method of Storch is to approximate e^{-s} by

$$H_n(s) = \frac{1}{M(s) + N(s)} \qquad \text{Bessel filter}$$

and associate the even and odd polynomials M and N with the even and odd functions $\cosh s$ and $\sinh s$ by

$$\frac{M(s)}{N(s)} = \frac{\cosh s}{\sinh s} \tag{7.45}$$

The hyperbolic functions are represented by their Maclaurin expansions,

$$\cosh s = 1 + \frac{s^2}{2!} + \frac{s^4}{4!} + \frac{s^6}{6!} + \cdots$$

$$\sinh s = s + \frac{s^3}{3!} + \frac{s^5}{5!} + \frac{s^7}{7!} + \cdots$$

and the right member of (7.45) is expressed in a Cauer 2 type of continued fraction expansion. The result is

(handwritten: Bessel filter)

(handwritten: |H|² = constant)

$$\frac{M(s)}{N(s)} = \frac{1}{s} + \cfrac{1}{\dfrac{3}{s} + \cfrac{1}{\dfrac{5}{s} + \cfrac{1}{\ddots \quad \cfrac{1}{\dfrac{2n-1}{s}}}}} \tag{7.46}$$

The odd integers, $1, 3, 5, \ldots$, appear as the coefficients of $1/s$, as shown. The process is truncated with n terms and the continued fraction is simplified to a rational function, the numerator taken as $M(s)$ and the denominator taken as $N(s)$.

As an example, suppose we wish a Bessel filter of order $n = 2$. By (7.46) we have

(handwritten: 1st   2nd)

$$\frac{M}{N} = \frac{1}{s} + \frac{1}{3/s} = \frac{s^2 + 3}{3s}$$

(handwritten: $H(s) = \dfrac{K}{M+N}$)

and thus the transfer function is

$$H(s) = \frac{K}{s^2 + 3s + 3} \tag{7.47}$$

(We have used a general constant K as the numerator rather than 1; quite often K is taken so that $H(0) = 1$, in which case $K = 3$ in this example.)

(handwritten: $H(0) = 1 = \dfrac{K}{0+3}$ ∴ $K=3$)

The time delay of (7.47) is given by

(handwritten: $T(\omega) = -\dfrac{d}{d\omega}\phi(\omega)$)

$$T(\omega) = -\frac{d}{d\omega}\left[-\tan^{-1}\frac{3\omega}{3 - \omega^2}\right]$$

$$= \frac{3\omega^2 + 9}{\omega^4 + 3\omega^2 + 9} \tag{7.48}$$

from which we note that

$$T(0) = 1, \qquad T(1) = 12/13 \tag{7.49}$$

This example was considered earlier in Exercise 1.2. As the reader may verify, using (7.46) etc., the third and fourth-order Bessel filter functions are given by

$$H(s) = \frac{15}{s^3 + 6s^2 + 15s + 15} \tag{7.50}$$

and

$$H(s) = \frac{105}{s^4 + 10s^3 + 45s^2 + 105s + 105} \tag{7.51}$$

with $T(0) = 1$ in both cases. In (7.50) we have

$$T(1) = 276/277 \tag{7.52}$$

and in (7.51) we have

$$T(1) = 12{,}745/12{,}746 \tag{7.53}$$

These last two examples were given earlier in Exercise 1.3.

Since constant time delay goes hand-in-hand with linear phase, we would expect the Bessel filter to have a remarkably good phase response. This is illustrated for the cases $n = 2, 3, 4,$ and 5 in Fig. 7.9. These may be compared with the Butterworth and Chebyshev phase responses of Fig. 3.14.

In general, the Bessel filter of order n is defined by the transfer function

$$H(s) = \frac{B_n(0)}{B_n(s)} \tag{7.54}$$

where $B_n(s)$ is obtained for a given n from the continued fraction expansion (7.46). The polynomials $B_n(s)$ may also be expressed as

$$B_n(s) = s^n y_n(1/s)$$

where $y_n(s)$ is a *Bessel* polynomial, so-called because of its relation to the spherical Bessel functions $J_{\pm(n+1/2)}(x)$ [KF]. From the known expression for the Bessel polynomials we may write

$$B_n(s) = \sum_{k=0}^{n} \frac{(2n - k)!\, s^k}{2^{n-k} k!\, (n-k)!}, \qquad n = 1, 2, 3, \ldots \tag{7.55}$$

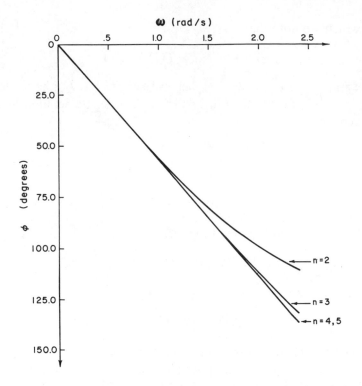

Figure 7.9. Bessel phase responses.

We note from (7.48) and Exercise 3.7 that the time delay for the second-order case is maximally flat. This is true in general, as Storch observed, and as Thomson in an earlier paper [Th] proved. (See also Exercise 7.5, for the cases $n = 1, 2$, and 3.)

As a final note on the Bessel filter, because of the relationship of $B_n(s)$ to the Bessel functions, it is possible to write [St]

$$B_n(j\omega) = e^{j\omega}\omega^{n+1}\sqrt{\frac{\pi}{2\omega}}[(-1)^n J_{-n-1/2}(\omega) - j J_{n+1/2}(\omega)] \qquad (7.56)$$

which can be used to find explicit expressions for the amplitude, phase, and time delay of the Bessel filter.

For the convenience of the reader, the polynomials $B_n(s)$ are given in Appendix A for $n = 1, 2, \ldots, 6$, and its second-order factors are given in Appendix B for $n = 2, 4$, and 6.

7.8 ALL-PASS FILTERS WITH
MAXIMALLY-FLAT TIME DELAY

Finally in this chapter, we consider a filter which combines the constant amplitude characteristic of the all-pass filter with the maximally-flat time delay of the Bessel filter. Such a filter is important if we want a constant time delay without changing the amplitude of the input signal. We shall consider only the normalized case, since the denormalization details are identical to those of the Bessel filter.

Budak [Bu-1] has given an approximation to the ideal normalized function e^{-s} by noting that

$$ e^{-s} = \frac{e^{-ks}}{e^{-(k-1)s}}, \qquad k > 0 \tag{7.57} $$

Using the approximations of the Bessel filter to e^{-s} given by (7.54),

$$ e^{-ks} = \frac{B_n(0)}{B_n(ks)}, \qquad n = 1, 2, 3, \dots $$

and

$$ e^{-(k-1)s} = \frac{B_m(0)}{B_m[(k-1)s]}; \qquad m = 1, 2, \dots ; m \leq n $$

we may write (7.57) in the form

$$ e^{-s} = \frac{B_n(0)B_m[(k-1)s]}{B_m(0)B_n(ks)} = H_{mn}^k(s) \tag{7.58} $$

which we have defined as $H_{mn}^k(s)$. We note that (7.58) reduces to the transfer function $H(s)$ of the Bessel filter when $k = 1$ and that $H(ks)$ results when $m = 0$. For $0 < k < 1$, (7.58) is a nonminimum phase function, and for $k \geq 1$ it is a minimum phase function.

It has recently been noted [MJJ] that if $m = n$, the time delay, T_{mn}^k, of (7.58) is maximally flat, and evidently if $k = 1/2$, then (7.58) is an all-pass function. Thus combining the two, $k = 1/2$ and $m = n$, we have an all-pass filter with maximally-flat time delay. Moreover, it is shown in reference [MJJ] that this case yields a flat response for approximately twice the frequency range of the Bessel filter. For example, if $T_2(\omega)$ is the time delay of the second-order normalized Bessel filter and

$T_{22}^{1/2}(\omega)$ is that of (7.58) ($m = n = 2$, $k = \frac{1}{2}$), then both are 1 at $\omega = 0$, but at $\omega = 1$ we have

$$T_2(1) = 12/13, \quad T_{22}^{1/2}(1) = 156/157 \qquad (7.59)$$

The first of (7.59) was shown earlier in connection with the Bessel filter, and the second follows from

$$H_{22}^{1/2}(s) = \frac{s^2 - 6s + 12}{s^2 + 6s + 12} \qquad (7.60)$$

for which

$$T_{22}^{1/2}(\omega) = \frac{144 + 12\omega^2}{144 + 12\omega^2 + \omega^4} \qquad (7.61)$$

This latter function is evidently maximally flat.

Another example is the third-order case for which

$$T_3(1) = 276/277, \quad T_{33}^{1/2}(1) = 15{,}144/15{,}145 \qquad (7.62)$$

To illustrate further the time-delay properties of $H_{nn}^{1/2}(s)$, Fig. 7.10 shows $T_{nn}^{1/2}(\omega)$ for $n = 1, 2, \ldots, 5$, compared with the third-order Bessel time delay.

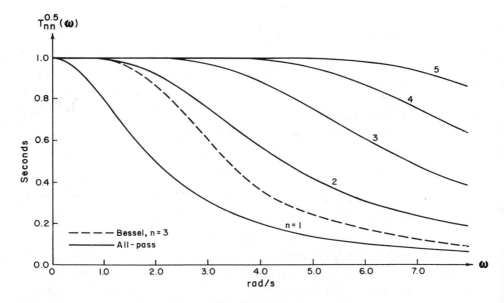

Figure 7.10. Time delay curves.

7.9 SUMMARY

All-pass filters have constant amplitude responses and nonminimum phase responses. One method of passively realizing all-pass functions, considered in this chapter, uses symmetrical lattices. Constant-resistance symmetrical lattices were also developed and were shown to be particularly useful, when cascaded, to realize higher-order transfer functions. In the case of all-pass functions, the constant-resistance lattices are LC networks.

The Bessel filter and an all-pass generalization of it were considered, and both were observed to approximate a maximally-flat time delay. Since these two filters are respectively of the low-pass and all-pass type, they may be realized by the methods considered in this and the previous chapter.

Up to this point most of our discussion of filters has been limited to the frequency domain. In the next chapter we shall consider time-domain characteristics and, in particular, the idea of distortion of the output signal.

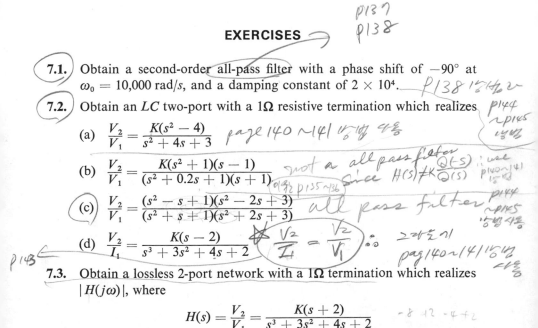

EXERCISES

7.1. Obtain a second-order all-pass filter with a phase shift of $-90°$ at $\omega_0 = 10{,}000$ rad/s, and a damping constant of 2×10^4.

7.2. Obtain an LC two-port with a 1Ω resistive termination which realizes

(a) $\dfrac{V_2}{V_1} = \dfrac{K(s^2 - 4)}{s^2 + 4s + 3}$

(b) $\dfrac{V_2}{V_1} = \dfrac{K(s^2 + 1)(s - 1)}{(s^2 + 0.2s + 1)(s + 1)}$

(c) $\dfrac{V_2}{V_1} = \dfrac{(s^2 - s + 1)(s^2 - 2s + 3)}{(s^2 + s + 1)(s^2 + 2s + 3)}$

(d) $\dfrac{V_2}{I_1} = \dfrac{K(s - 2)}{s^3 + 3s^2 + 4s + 2}$

7.3. Obtain a lossless 2-port network with a 1Ω termination which realizes $|H(j\omega)|$, where

$$H(s) = \frac{V_2}{V_1} = \frac{K(s + 2)}{s^3 + 3s^2 + 4s + 2}$$

7.4. Obtain Bessel filters of orders 3 and 4 as LC two-ports terminated in a 1Ω resistance, which approximate $T(1) = 1$s. Frequency and impedance scale the networks so that the terminating resistance is $1\ k\Omega$ and at $\omega_0 = 10{,}000\ rad/s$ we have $T \approx 100\ \mu s$.

7.5. Show that if T_k is the time delay for the normalized Bessel filter of order k, then

$$T_1 = \frac{1}{1 + \omega^2}$$

$$T_2 = \frac{9 + 3\omega^2}{9 + 3\omega^2 + \omega^4}$$

$$T_3 = \frac{225 + 45\omega^2 + 6\omega^4}{225 + 45\omega^2 + 6\omega^4 + \omega^6}$$

Note that by Exercise 3.7, these are all maximally flat.

7.6. Obtain $T_{33}^{1/2}(\omega)$ and verify the second of (7.62).

7.7. Obtain a singly-terminated LC realization of (7.60) and scale the result so that the time delay at $\omega_0 = 10{,}000\ rad/s$ is approximately $100\ \mu s$ and the terminating resistance is $1\ k\Omega$.

7.8. Verify that the odd integers appear in the continued fraction expansion (7.46) for $n = 6$.

CHAPTER

8

Time Domain
Considerations

8.1 SIGNAL DISTORTIONS

In the previous chapters, except for a brief discussion of time delay in Chapter 7, we have restricted ourselves to the frequency domain, considering the transfer functions $H(s)$ or $H(j\omega)$. In this chapter we shall briefly discuss some time-domain characteristics of the input and output signals, $v_1(t)$ and $v_2(t)$, particularly the *signal distortions*. For a more thorough treatment the reader is referred to such works as [RB], [H], [K], [P-2], and [S-1].

We begin by recalling from Chapter 7 that the output signal $v_2(t)$ is defined to be undistorted if

$$v_2(t) = Kv_1(t - \tau) \tag{8.1}$$

where K and τ are constants, defined respectively as the *gain* and *delay*. The required (ideal) transfer function in this case is

$$H(s) = Ke^{-s\tau} \tag{8.2}$$

The ideal transfer function is unrealizable, of course, since it is not a rational function; thus in any realizable system the output signal is distorted in various ways. In particular, in the frequency domain, the departures from ideal are (1) the amplitude $|H(j\omega)|$ is not constant, (2) the phase $\phi(\omega)$ is not linear, and (3) the time delay $T(\omega)$ is not constant. In the time domain, the distortions appear in the output signal, $v_2(t) = \mathcal{L}^{-1}V_2(s)$, or

$$v_2(t) = \mathcal{L}^{-1}H(s)V_1(s) \tag{8.3}$$

Two outputs which are often used to measure distortion are the *impulse response*

$$h(t) = \mathcal{L}^{-1}H(s) \tag{8.4}$$

which is the response to an impulse input, $v_1(t) = \delta(t)$ $(V_1(s) = 1)$, and the *step response*

$$a(t) = \mathcal{L}^{-1}[H(s)/s] \tag{8.5}$$

In the latter case the input is the unit step function $u(t)$, defined by

$$u(t) = 0, \qquad t < 0$$
$$= 1, \qquad t > 0 \tag{8.6}$$

and having a transform $\mathcal{L}u(t) = 1/s$.

We shall limit ourselves to a brief consideration of the impulse and step responses in the case of low-pass filters.

8.2 THE STEP AND IMPULSE RESPONSES

The ideal step response is given by (8.5) when the transfer function is the ideal function of (8.2). That is,

$$a(t) = \mathcal{L}^{-1}[Ke^{-s\tau}/s]$$
$$= Ku(t - \tau) \tag{8.7}$$

shown for $K = 1$ in Fig. 8.1. Evidently $a(t)$ is simply a step function delayed by τ seconds.

The departures from ideal are shown in Fig. 8.2, which is a typical step response of a realizable system. To discuss the deviations from ideal we shall find it convenient to define a number of figures of merit, some of which are shown in the figure. First, the phenomenon of *ringing* is the oscillatory transient shown between t_2 and t_s in Fig. 8.2.

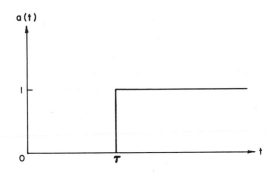

Figure 8.1. An ideal step response.

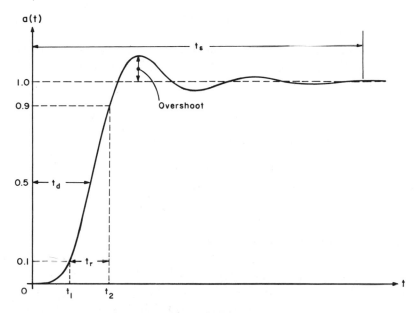

Figure 8.2. A realizable step response.

Ringing is the result of a sudden change in the input signal, such as, in this case, a step function. The *settling time* t_s is the time beyond which the step response remains within some specified range, usually taken as $\pm 2\%$, of its final value. *Overshoot* is the difference between the peak value and the final value of the step response, expressed as a percentage of the final value. Thus ringing may be measured quantitatively by the overshoot and the settling time. There is also, in some cases, an *undershoot*, when the signal is oscillatory at first before starting its rise.

Two other figures of merit shown in Fig. 8.2 are the *rise time* t_r and the *delay time* t_d. As indicated, $t_r = t_2 - t_1$ is the time required for the step response to rise from 10% to 90% of its final value, and t_d is the time required for the response to reach 50% of its final value.

Comparing Figs. 8.1 and 8.2 we see that ideally there is no ringing, $t_s = t_d = \tau$, and $t_r = 0$. There is, of course, also no overshoot nor undershoot.

In the case of low-pass filters, there is a relationship between the rise time t_r in the time domain and the bandwidth of the frequency domain [RB]. As long as the overshoot is small, say less than 5%, for most practical filters the rise time-bandwidth product is given approximately by

$$t_r \omega_c = 2.2 \tag{8.8}$$

As a simple example, let us consider the low-pass RC filter of Exercise 1.5 for the case $RC = 1$. The transfer function is

$$H(s) = \frac{1}{s+1}$$

and $\omega_c = 1$ rad/s. The step response is given by

$$a(t) = \mathcal{L}^{-1}\left[\frac{1}{s(s+1)}\right]$$
$$= (1 - e^{-t})u(t)$$

Thus the rise time is given by $t_r = t_2 - t_1$, where

$$1 - e^{-t_2} = 0.9$$

and

$$1 - e^{-t_1} = 0.1$$

Solving these equations yields $t_2 = 2.3026\ s$ and $t_1 = 0.1054\ s$, which results in $t_r = 2.1972\ s$. Since $\omega_c = 1$ rad/s, we have

$$t_r \omega_c = 2.1972$$

The ideal impulse response is given by

$$h(t) = \mathcal{L}^{-1}Ke^{-s\tau}$$
$$= K\delta(t - \tau)$$

which is an impulse of "magnitude" K delayed τ seconds. In a practical circuit this will be approximated by a response somewhat like that of Fig. 8.3.

Figure 8.3. A realizable impulse response.

For example, suppose we have the second-order low-pass transfer function

$$H(s) = \frac{K}{(s+a)^2 + b^2}$$

The inverse transform is given by

$$h(t) = \frac{K}{b} e^{-at} \sin bt$$

whose sketch resembles Fig. 8.3.

In the case of ideal low-pass filters (ideal low-pass amplitude and linear phase), it may be shown [H] that the step and impulse responses are similar to the responses of Figs. 8.2 and 8.3 respectively, except that there is undershoot in both cases and the responses are not zero for $t < 0$ (so-called *noncausal* circuit responses). We shall consider these responses for the Butterworth, Chebyshev, and Bessel filters in the next section.

8.3 LOW-PASS FILTER RESPONSES

Overshoot is generally caused by high gains at the higher frequencies, and ringing is due to sharp cutoff in the amplitude response [K]. Thus we should expect more ringing and longer settling time as the amplitude response approaches more closely the ideal low-pass amplitude response. This happens in the case of both Butterworth and Chebyshev low-pass filters, because as the order n increases the amplitude increases for frequencies in the passband near the cutoff point. Also, of course, the cutoff is sharper as n increases.

On the other hand, the Bessel amplitude and phase responses approach the ideal of (8.2) as n increases. Thus the higher the order the less the ringing and the shorter the settling time. In the case of the Butterworth and Chebyshev filters, higher n is accompanied by greater rise times, but in the Bessel case, the rise time decreases as n increases. Of course, for the Bessel filter, the delay time should vary little with n.

To illustrate these properties, we have depicted step responses of various orders of Butterworth, Chebyshev (for $\epsilon = 0.5$), and Bessel filters in Figs. 8.4, 8.5, and 8.6 respectively.

In the case of the impulse response, we should expect the Bessel filter to better approximate the ideal than either the Butterworth or the Chebyshev filters. That is, the impulse response of the Bessel should more nearly approximate a single pulse at $\omega = 1$ rad/s (normalized case) with very little undershoot. The Butterworth and Chebyshev

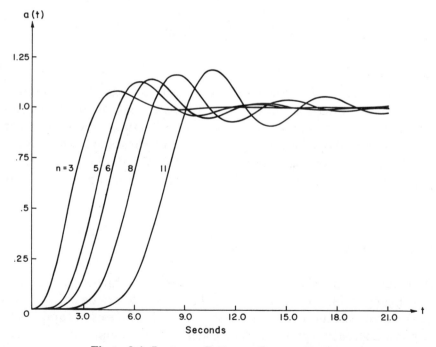

Figure 8.4. Low-pass Butterworth step responses.

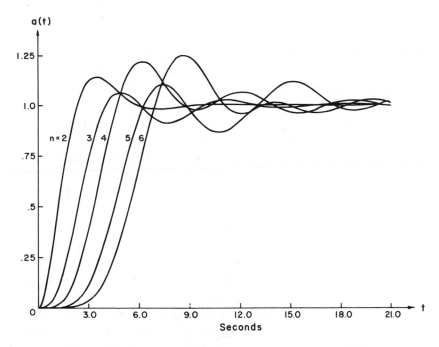

Figure 8.5. Low-pass Chebyshev step responses ($\epsilon = 0.5$).

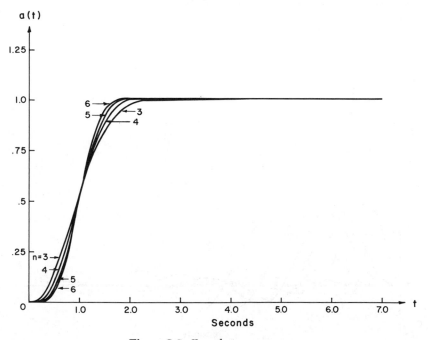

Figure 8.6. Bessel step responses.

161

responses should be pulses centered at various points, $t = t_0$, as n changes, with an oscillating transient (undershoot) for $t > t_0$. These characteristics are illustrated in Figs. 8.7, 8.8, and 8.9.

The decision as to which filter is best depends upon which characteristics are important and which are relatively unimportant. For example, if sharpness of cutoff is the dominant factor, then for a given order a Chebyshev filter is to be preferred over a Butterworth. However, the Chebyshev filter, because of its sharper cutoff, has more ringing and a longer settling time than the Butterworth or the Bessel filters, which could be an objectionable feature. In particular, in transmitting a sequence of pulses, filters with long rise and settling times are undesirable because the output pulses "smear" over each other and tend to be indistinguishable [S-1]. Also, there are systems which may be damaged by excessive overshoot, in which case sharper cutoff must be sacrificed for better time-domain characteristics.

In case overshoot minimization and good amplitude response are both important, consideration might be given to using the inverse Chebyshev filter. As was noted in Chapter 6, it compares favorably with the Chebyshev filter, but it has a flat passband response, which should yield a better time-domain response.

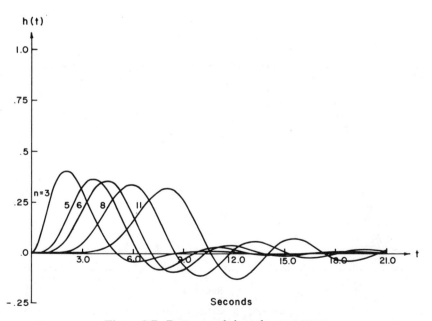

Figure 8.7. Butterworth impulse responses.

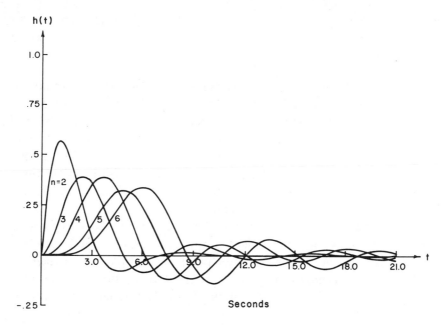

Figure 8.8. Chebyshev impulse responses ($\epsilon = 0.5$).

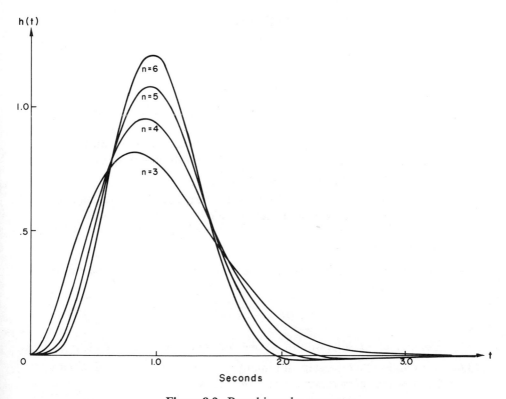

Figure 8.9. Bessel impulse responses.

163

8.4 SUMMARY

An input signal is undistorted in passing through the filter if it is changed only by amplification and/or by being delayed. In this chapter distortion was considered for a low-pass filter by studying its impulse and step responses and comparing them with the ideal cases. In general, filters with better amplitude responses, such as Chebyshev filters, exhibit inferior time-domain responses, while those with excellent time-delay features (and poor low-pass amplitudes), such as Bessel filters, have good time-domain characteristics.

In considering which filter should be used, it may be necessary to compromise between an excellent amplitude response and excellent time-domain characteristics. In some cases, as will be seen in Exercise 8.7, it may be possible to compensate for a network's poor time-domain responses and still retain some of its desirable frequency-domain responses.

EXERCISES

8.1. Find the step and impulse responses for a second-order normalized low-pass Butterworth filter with a gain of 1.

8.2. Repeat Exercise 8.1 for the third-order case.

$$\text{Ans.:}\ a(t) = \left[1 - e^{-t} - \frac{2}{\sqrt{3}}e^{-2t}\sin\frac{\sqrt{3}\,t}{2}\right]u(t)$$

8.3. Find the step and impulse responses for a second-order normalized Bessel filter with a gain of 1. Compare the results with those of Exercise 8.1 by sketching $a(t)$ in both cases.

8.4. Find the step and impulse responses for the all-pass filter having maximally-flat time delay, whose transfer function is given by (7.60). Note that $a(0) = 1$, and compare the results with those of Exercise 8.3 by sketching $a(t)$.

8.5. Find the step and impulse responses for the all-pass filter whose transfer function is given by

$$H(s) = \frac{s^2 - 2s + 2}{s^2 + 2s + 2}$$

Compare the results with those of Exercise 8.4 by sketching $a(t)$.

8.6. Find and sketch the step and impulse responses for the bandpass and band-reject filters whose transfer functions are given by

(a) $H(s) = \dfrac{s}{s^2 + 0.2s + 1}$

(b) $H(s) = \dfrac{s^2 + 1}{s^2 + 0.2s + 1}$

8.7. For the network of Exercise 1.5, $t_r \approx 2.2/\omega_c = 2.2\,RC$ and the gain is 1. Show that by compensating the network with the arrangement using a controlled source as shown in Fig. Ex. 8.7, the rise time may be shortened. Determine K so that the gain is still 1.

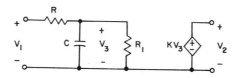

Figure Ex. 8.7.

$$\text{Ans.: } K = 1 + \frac{R}{R_1}, \qquad t_r \approx 2.2\,RC\left(\frac{R_1}{R + R_1}\right)$$

9

Passive Synthesis—
Further Considerations

9.1 DOUBLY-TERMINATED REALIZATIONS

In Chapter 5 we considered realizations of a given transfer function with a lossless two-port network terminated in a load resistor. A more general realization is a *doubly-terminated* network consisting of a lossless two-port network terminated in a resistor R_1 at the source port and in a resistor R_2 at the load port, as shown in Fig. 9.1. Such a network is capable of providing greater power flow in the passband with the consequence that the circuit is less sensitive at passband frequencies to change in its element values [TM]. The synthesis of the doubly-terminated lossless network was given by Darlington [D], requiring ideal trans-

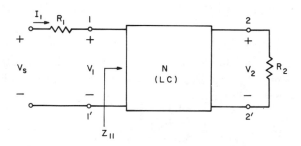

Figure 9.1. A doubly-terminated *LC* network.

formers in the general case. We shall restrict ourselves here to the case of all-pole transfer functions, for which a ladder realization is always possible. The development we give is somewhat similar to that of Weinberg [W].

The circuit of Fig. 9.1 may be thought of as an *LC* network *inserted* between the input and output resistors. If P_2 is the power delivered to the load with the network inserted, and P_{20} is the power that would be delivered to the load without the inserted network, as shown in Fig. 9.2

Figure 9.2. The system without the inserted network.

$(P_{20} = |V_{20}(j\omega)|^2/2R_2)$, then some authors define an *insertion loss* proportional to $\log (P_{20}/P_2)$. The procedure for finding the inserted network for a specified set of conditions is known as *insertion loss synthesis*.

Referring to Fig. 9.1, we shall be interested in the transfer function

$$than\ point\ 1-1'\ terminal\ with\ voltage\ source\quad H_s = G_{2s} = \frac{V_2(s)}{V_s(s)} \tag{9.1}$$

where V_s is the source voltage. (If we wish to consider a current source I_s as input, we may do so by replacing the network to the left of the 1–1' terminals by its Norton equivalent, a shunt resistor R_1 in parallel with the current source $I_s = V_s/R_1$.) The impedance $Z_{11} = V_1/I_1$ is the input impedance at the 1–1' port with the 2–2' port terminated in R_2.

The *available* power P_A is defined as the maximum power the source V_s, with internal resistance R_1, can deliver. By the maximum power transfer theorem, P_A is delivered to the load R_2 when in Fig. 9.2, $R_2 = R_1$. Therefore we have

$$P_A = \frac{|V_s(j\omega)|^2}{8R_1} \tag{9.2}$$

The power P_L delivered to the load in Fig. 9.1 is evidently

$$P_L = \frac{|V_2(j\omega)|^2}{2R_2} \tag{9.3}$$

which cannot exceed P_A.

We define the *transmission coefficient $t(s)$* by the relation

$$|t(j\omega)|^2 = \frac{P_L}{P_A} \tag{9.4}$$

which by (9.1), (9.2), and (9.3) may be written

$$|t(j\omega)|^2 = \frac{4R_1}{R_2}|G_{2s}(j\omega)|^2 \tag{9.5}$$

The *reflection coefficient $\rho(s)$* is defined by

$$|\rho(j\omega)|^2 = 1 - |t(j\omega)|^2 \tag{9.6}$$

Since by (9.4), $|t(j\omega)|^2$ is the fraction of the available power that is absorbed by the load, then $|\rho(j\omega)|^2$ is the fraction that is rejected or *reflected* by the load. In the passive case,

$$|t(j\omega)|^2 \leq 1 \tag{9.7}$$

which must also hold for $|\rho(j\omega)|^2$.

If we define $Z_{11}(j\omega)$ by

$$Z_{11}(j\omega) = R_{11} + jX_{11}$$

then the power delivered to the 1–1' terminals in Fig. 9.1 is given by $R_{11}|I_1(j\omega)|^2/2$, and since the two-port network is lossless, this must also be the power delivered to the load. Therefore we have, by (9.3),

$$R_{11}|I_1(j\omega)|^2 = \frac{|V_2(j\omega)|^2}{R_2} \tag{9.8}$$

Also from Fig. 9.1 we see that

$$I_1 = \frac{V_s}{R_1 + Z_{11}}$$

so that (9.8) becomes

$$\frac{R_{11}|V_s(j\omega)|^2}{|R_1 + Z_{11}(j\omega)|^2} = \frac{|V_2(j\omega)|^2}{R_2}$$

This may be written

$$|G_{2s}(j\omega)|^2 = \left|\frac{V_2(j\omega)}{V_s(j\omega)}\right|^2 = \frac{R_2 R_{11}}{|R_1 + Z_{11}(j\omega)|^2}$$

which substituted into (9.5) yields

$$|t(j\omega)|^2 = \frac{4R_1 R_{11}}{|R_1 + Z_{11}(j\omega)|^2}$$

Substituting this expression into (9.6) we have

$$\begin{aligned}
|\rho(j\omega)|^2 &= 1 - \frac{4R_1 R_{11}}{|R_1 + Z_{11}(j\omega)|^2} \\
&= \frac{|R_1 + Z_{11}(j\omega)|^2 - 4R_1 R_{11}}{|R_1 + Z_{11}(j\omega)|^2} \\
&= \frac{(R_1 + R_{11})^2 + X_{11}^2 - 4R_1 R_{11}}{|R_1 + Z_{11}(j\omega)|^2} \\
&= \frac{R_1^2 - 2R_1 R_{11} + R_{11}^2 + X_{11}^2}{|R_1 + Z_{11}(j\omega)|^2} \\
&= \frac{(R_1 - R_{11})^2 + X_{11}^2}{|R_1 + Z_{11}(j\omega)|^2}
\end{aligned}$$

or finally

$$|\rho(j\omega)|^2 = \frac{|R_1 - Z_{11}(j\omega)|^2}{|R_1 + Z_{11}(j\omega)|^2} \tag{9.9}$$

By the results of Sec. 2.5 we may then obtain

$$\rho(s) = \pm\frac{R_1 - Z_{11}(s)}{R_1 + Z_{11}(s)} \tag{9.10}$$

We note that in the process of obtaining (9.10) we have two possible values of $\rho(s)$. Also we should select a Hurwitz denominator, as in the case of transfer functions, since from (9.5) we conclude that $t(s)$ has the same Hurwitz denominator as $G_{2s}(s)$, and from (9.6) we conclude that $p(s)$ will also have this denominator.

Solving (9.10) for $Z_{11}(s)$ we see that there are two possible answers, given by

$$Z_{11}(s) = R_1 \frac{1 - p(s)}{1 + p(s)} \tag{9.11}$$

and

$$Z_{11}(s) = R_1 \frac{1 + p(s)}{1 - p(s)} \tag{9.12}$$

It may be shown [W] that if $|t(j\omega)|^2 \le 1$, then $Z_{11}(s)$ is passively realizable as a driving-point impedance. Realization of Z_{11} in either form leads to a realization of the given transfer function.

Thus the insertion loss method of synthesis consists in determining $|t(j\omega)|^2$ from a given $G_{2s}(s)$, and finding in turn $|p(j\omega)|^2$, $p(s)$, and $Z_{11}(s)$, all for a given R_1 and R_2. Then the realization of $Z_{11}(s)$ yields the completed synthesis. In the next section we consider these steps for a special, but very important case.

9.2 DOUBLY-TERMINATED LADDERS

In this section we shall consider what values of the terminating resistors R_1 and R_2 are permissible in the double termination realization of Fig. 9.1, for the special case

$$G_{2s}(s) = \frac{V_2(s)}{V_s(s)} = K\frac{P(s)}{Q(s)}$$

$$= K\frac{s^m + a_{m-1}s^{m-1} + \cdots + a_0}{s^n + b_{n-1}s^{n-1} + \cdots + b_0} \tag{9.13}$$

where $a_0, b_0 \ne 0$. That is, G_{2s} has no pole or zero at $s = 0$. Later we shall restrict G_{2s} further to the all-pole case and obtain doubly-terminated ladder realizations.

From (9.5) we may write

$$t(s) = 2\sqrt{\frac{R_1}{R_2}} G_{2s}(s) \tag{9.14}$$

or

$$t(s) = 2K\sqrt{\frac{R_1}{R_2}} \frac{P(s)}{Q(s)} \tag{9.15}$$

Since G_{2s} has no transmission zero at $s = 0$, any ladder realization can have at $s = 0$ no open-circuit series element or short-circuit shunt element. Since the elements are LC elements, if at $s = 0$ they are not open-circuits then they must be short-circuits, and conversely, if they are not short-circuits, then they must be open-circuits. Therefore at $s = 0$, the circuit looks like that of Fig. 9.2, so that

$$G_{2s}(0) = \frac{R_2}{R_1 + R_2} \tag{9.16}$$

Also from (9.13) we have

$$G_{2s}(0) = K\frac{P(0)}{Q(0)} = \frac{Ka_0}{b_0}$$

so that

$$K = \frac{R_2 b_0}{(R_1 + R_2)a_0} = \frac{R_2}{(R_1 + R_2)P(0)/Q(0)} \tag{9.17}$$

By (9.13), (9.15), and (9.17) we may write

$$|t(j\omega)| = 2\sqrt{\frac{R_1}{R_2}}K\left|\frac{P(j\omega)}{Q(j\omega)}\right|$$

$$= \frac{2\sqrt{R_1 R_2}|P(j\omega)/Q(j\omega)|}{(R_1 + R_2)|P(0)/Q(0)|} \tag{9.18}$$

Therefore $|t(j\omega)| \leq 1$, as required, if

$$\left|\frac{P(j\omega)}{Q(j\omega)}\right|_{max} \leq \frac{R_1 + R_2}{2\sqrt{R_1 R_2}}\left|\frac{P(0)}{Q(0)}\right| \tag{9.19}$$

where the subscript "max" refers to the maximum value of $|P(j\omega)/Q(j\omega)|$. Equation (9.19) is equivalent to

$$\frac{2\sqrt{R_1 R_2}}{R_1 + R_2} \leq \frac{|G_{2s}(0)|}{|G_{2s}(j\omega)|_{max}} = \frac{|t(0)|}{|t(j\omega)|_{max}} \tag{9.20}$$

Therefore we see that if $|G_{2s}(j\omega)|_{max}$ (or $|t(j\omega)|_{max}$) does not occur at $\omega = 0$, then

$$\frac{2\sqrt{R_1 R_2}}{R_1 + R_2} < 1$$

and hence we cannot have $R_1 = R_2$. (For $R_1 = R_2$ yields $2\sqrt{R_1 R_2}/(R_1 + R_2) = 1$.)

Let us consider now the all-pole function

$$G_{2s}(s) = \frac{Ka_0}{Q(s)} = \frac{Ka_0}{s^n + b_{n-1}s^{n-1} + \cdots + b_0} \tag{9.21}$$

for which we have, by (9.15) and (9.17),

$$t(s) = \frac{2\sqrt{R_1 R_2}b_0/(R_1 + R_2)}{Q(s)} \tag{9.22}$$

and consequently

$$|t(j\omega)| = \frac{2\sqrt{R_1 R_2}b_0/(R_1 + R_2)}{|Q(j\omega)|} \tag{9.23}$$

Thus, since $|t(j\omega)| \le 1$, we must have

$$|t(j\omega)|_{max} = \frac{2\sqrt{R_1 R_2}b_0/(R_1 + R_2)}{|Q(j\omega)|_{min}} \le 1$$

$\left(|Q(j\omega)|_{min} = 1\right)$

$k = \dfrac{2\sqrt{R_1 R_2}}{R_1 + R_2}$

or

$$\frac{2\sqrt{R_1 R_2}}{R_1 + R_2} \le \frac{|Q(j\omega)|_{min}}{b_0} \tag{9.24}$$

In summary, to construct a doubly-terminated LC ladder with a given all-pole function $G_{2s}(s)$, $|G_{2s}(j\omega)|$, or $|t(j\omega)|$, we determine suitable values of R_1 and R_2 from (9.24), obtain $|t(j\omega)|$ or $t(s)$ from (9.22), obtain $p(s)$ from (9.6), obtain $Z_{11}(s)$ from (9.11) or (9.12), and obtain a continued fraction expansion about infinity for $Z_{11}(s)$. Interpreting this expansion leads to the completed network. There are two possibilities since there are two possible Z_{11} expressions. One leads to the correct R_2 and the other yields $1/R_2$ as the load resistor. The difference is in interpreting the continued fraction expansion, inasmuch as one begins with an impedance and the other with an admittance. Indeed, if $R_1 = 1\Omega$, the two expressions for Z_{11} are reciprocals and either yields the correct termination. Finally, if $|G_{2s}(j\omega)|$ (or $|t(j\omega)|$) takes on its maximum value at $\omega = 0$, it is possible to take $R_1 = R_2$, as may be seen from (9.20).

The transfer function we obtain by this process is given by (9.14) and (9.22) to be

$$G_{2s}(s) = \frac{V_2(s)}{V_s(s)} = \frac{\left[\dfrac{R_2 b_0}{R_1 + R_2}\right]}{Q(s)} \tag{9.25}$$

Extensive tables of normalized, doubly-terminated ladder network elements have been compiled for Butterworth, Chebyshev, Bessel, and elliptic filters. The first three are well-documented by Weinberg [W] for various combinations of terminating resistances, and in the Chebyshev case, for various ripple widths. Saal and Ulbrich [SU] have compiled the network elements for various values of passband and stopband ripples in the case of the elliptic low-pass filter. Their work is reproduced elsewhere, such as by Geffe [Ge-1], for example.

9.3 EXAMPLES OF DOUBLY-TERMINATED LADDERS

As an example of the double termination procedure, suppose we are given

$$|t(j\omega)|^2 = \frac{1}{1 + \omega^6} \tag{9.26}$$

From this we obtain

$$|p(j\omega)|^2 = 1 - \frac{1}{1 + \omega^6}$$

$$= \frac{\omega^6}{1 + \omega^6}$$

and

$$|p(j\omega)|^2 = p(s)p(-s) = \left.\frac{\omega^6}{1 + \omega^6}\right|_{\omega^2 = -s^2}$$

$$= \frac{-s^6}{1 - s^6}$$

The denominator $1 - s^6$ was previously considered in Sec. 2.5 and shown to have factors $Q(s)$ and $Q(-s)$, where

$$Q(s) = s^3 + 2s^2 + 2s + 1$$

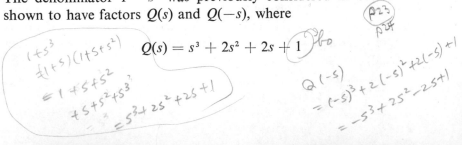

Thus we have, choosing s^3 rather than $-s^3$ as the numerator (either is acceptable),

$$p(s) = \frac{s^3}{s^3 + 2s^2 + 2s + 1}$$

(9.27)

[handwritten: $a_0 = \not\!\!p$ there is no constant]

[handwritten: b_0]

[handwritten: numerator $a_0 \neq 0$ $b_0 \neq 0$]

and by (9.22), since $b_0 = 1$, we have

$$t(s) = \frac{2\sqrt{R_1 R_2}/(R_1 + R_2)}{s^3 + 2s^2 + 2s + 1}$$

(9.28)

[handwritten: CASE 1> WHEN $R_1 = R_2$]

Evidently $|t(j0)| = |t(j\omega)|_{\max}$ by (9.26), so that $R_1 = R_2$ is possible. In this case the numerator of $t(s)$ in (9.28), and hence of $|t(j\omega)|^2$, is 1. Thus no scaling of $|t(j\omega)|^2$ is required. Choosing $R_1 = R_2 = 1\Omega$, we have by (9.11), (9.12), and (9.27), the two possible expressions

$$Z_{11}(s) = \frac{1 - \dfrac{s^3}{s^3 + 2s^2 + 2s + 1}}{1 + \dfrac{s^3}{s^3 + 2s^2 + 2s + 1}}$$

$$= \frac{2s^2 + 2s + 1}{2s^3 + 2s^2 + 2s + 1}$$

(9.29)

and the reciprocal expression,

$$Z_{11} = \frac{2s^3 + 2s^2 + 2s + 1}{2s^2 + 2s + 1}$$

(9.30)

Choosing (9.30), we have the continued fraction expansion

$$2s^2 + 2s + 1 \,)\, 2s^3 + 2s^2 + 2s + 1 \,(\, s \longleftarrow Z$$
$$\underline{2s^3 + 2s^2 + \ \ s}$$
$$s + 1 \,)\, 2s^2 + 2s + 1 \,(\, 2s \longleftarrow Y$$
$$\underline{2s^2 + 2s}$$
$$1 \,)\, s + 1 \,(\, s \longleftarrow Z$$
$$\underline{s}$$
$$1 \,)\, 1 \,(\, 1 \longleftarrow Y$$
$$\underline{1}$$

(9.31)

The network with the required $R_1 = 1\Omega$ is shown in Fig. 9.3. We note that $G_{2s}(s)$ for this case is given by (9.25) to be

$$G_{2s}(s) = \frac{\frac{1}{2}}{s^3 + 2s^2 + 2s + 1}$$

This is the same transfer function and Fig. 9.3 is the same circuit that we considered in Sec. 2.4.

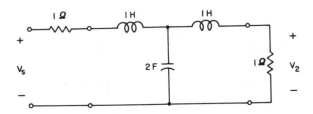

Figure 9.3. A realization of Eq. (9.26).

A dual development results if we use the Z_{11} of (9.29). In this case, since $R_1 = 1\Omega$, we need only interpret the continued fraction (9.31) as that of an admittance. That is, the first element is a Y, the next a Z, etc. The result is shown in Fig. 9.4.

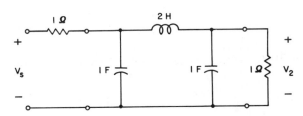

Figure 9.4. An alternate realization of Eq. (9.26).

Both realizations of (9.26) have the required terminations $R_1 = R_2 = 1\Omega$. In general however, if $R_2 \neq 1$, then one realization will have a termination R_2, but the other will have a termination $1/R_2$. If $R_1 = 1$, both realizations may be obtained from the same continued fraction, so it will be evident which is the desired one.

Let us now illustrate the case $R_1 \neq R_2$ by the example of (9.26). As in the previous example, we have $Q(s) = s^3 + 2s^2 + 2s + 1$ and thus $b_0 = 1$. Since $|Q(j\omega)|_{min} = 1$, (9.24) becomes

$$\frac{2\sqrt{R_1 R_2}}{R_1 + R_2} \leq 1 \qquad (9.32)$$

(handwritten: $\frac{2\sqrt{R_1 R_2}}{R_1 + R_2} \leq \frac{|Q(j\omega)|_{min}}{b_0}$ *)*

(handwritten: CASE 2 $R_1 \neq R_2$)

and thus we could let $R_1 = R_2$ as in the previous example. However, let us choose $R_1 = 1\Omega$ and $R_2 = 2\Omega$, in which case (9.32) is still valid. By (9.23) we see that $|t(j\omega)|^2$ must be scaled to

(handwritten: $\left(\frac{2\sqrt{R_1 R_2}}{R_1 + R_2}\right)^2 = \left(\frac{2\sqrt{1\cdot 2}}{1+2}\right)^2 = \left(\frac{2\sqrt{2}}{3}\right)^2$ *)*

$$|t(j\omega)|^2 = \frac{8/9}{1 + \omega^6}$$

and thus we have

(handwritten: $|t(j\omega)| = \frac{2\sqrt{R_1 R_2}\, b_0 / (R_1 + R_2)}{|Q(j\omega)|}$ *)*

$$|p(j\omega)|^2 = \frac{\frac{1}{9} + \omega^6}{1 + \omega^6}$$

(handwritten: $|p(j\omega)|^2 = 1 - |t(j\omega)|^2 = 1 - \frac{\frac{8}{9}}{1+\omega^6} = \frac{1 + \omega^6 - \frac{8}{9}}{1+\omega^6}$ *)*

This leads to

(handwritten: $p(s)p(-s) = \left.\frac{\frac{1}{9}+\omega^6}{1+\omega^6}\right|_{\omega^2 = -s^2}$ *)*

$$p(s)p(-s) = \frac{\frac{1}{9} - s^6}{1 - s^6}$$

which we may write as

(handwritten: $\frac{1}{9} - (s^3)^2 = (\frac{1}{3} - s^3)(\frac{1}{3} + s^3)$; $(\frac{1}{3})^2 - (s^3)^2$ *)*

$$p(s)p(-s) = \frac{\left(s^3 + \frac{1}{3}\right)\left(-s^3 + \frac{1}{3}\right)}{(s^3 + 2s^2 + 2s + 1)(-s^3 + 2s^2 - 2s + 1)}$$

(We do not need a Hurwitz numerator so there is no need to factor the numerator further.) Choosing

$$p(s) = \frac{s^3 + \frac{1}{3}}{s^3 + 2s^2 + 2s + 1}$$

we have from (9.11) and (9.12),

(handwritten: $p(-s) = \frac{-s^3 + \frac{1}{3}}{-s^3 + 2s^2 - 2s + 1}$ *)*

$$Z_{11} = \frac{2s^2 + 2s + \frac{2}{3}}{2s^3 + 2s^2 + 2s + \frac{4}{3}}$$

(handwritten: $Z_{11}(s) = R_1 \frac{1 - p(s)}{1 + p(s)}$ *)*

(handwritten: or $Z_{11}(s) = R_1 \frac{1 + p(s)}{1 - p(s)}$ *)*

(handwritten: $Z_{11} = \frac{1}{Z_{11}(s)}$ (reciprocal of $Z_{11}(s)$) *)*

or its reciprocal. The continued fraction expansion is

$$2s^2 + 2s + 2/3 \overline{)\, 2s^3 + 2s^2 + \;\; 2s + 4/3 \,}(s$$
$$\underline{2s^3 + 2s^2 + \;\; 2s/3}$$
$$\tfrac{4}{3}s + \tfrac{4}{3} \;\overline{)\, 2s^2 + 2s + 2/3\,}(\tfrac{3}{2}s$$
$$\underline{2s^2 + 2s}$$
$$2/3\,\overline{)\,\tfrac{4}{3}s + \tfrac{4}{3}\,}(\; 2s$$
$$\underline{\tfrac{4}{3}s}$$
$$\tfrac{4}{3}\,\overline{)\,\tfrac{2}{3}\,}(\tfrac{1}{2}$$
$$\underline{\tfrac{2}{3}}$$

Since we want $R_2 = 2\Omega$ and can use either Z_{11} or its reciprocal, we choose the case where the last quotient $\tfrac{1}{2}$ in the expansion is a Y. This requires the first quotient to be a Z, the second to be a Y, etc., resulting in the network of Fig. 9.5. We note from (9.25) that the transfer func-

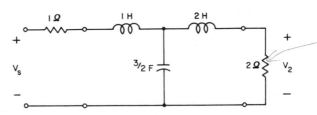

Figure 9.5. Another alternate realization of Eq. (9.26).

tion we have realized is

$$G_{2s} = \frac{2/3}{s^3 + 2s^2 + 2s + 1}$$

$$\frac{R_2 b_0}{R_1 + R_2} = \frac{z(1)}{1+2}$$

9.4 PREDISTORTION

In the synthesis procedures we have considered thus far the inductors and capacitors have been assumed to be lossless elements. That is, there is no *dissipation* associated with them. The actual devices are, of course, not lossless. Because a typical inductor is made of one or more turns of wire, it has resistance, and since the dielectric material of a capacitor is not a perfect insulator, it has finite resistance.

The *lossy* nature of the nonideal inductors and capacitors may be measured by the quantities q_L and q_C defined by

$$q_L = \frac{\omega L}{R_L} \tag{9.33}$$

and

$$q_c = \omega R_c C \tag{9.34}$$

where R_L and R_C are the resistances associated with the nonideal inductance L and capacitance C, respectively. It may be possible to approximate q_L and q_C by constants, measured at some point $\omega - \omega_0$, where ω_0 is the frequency of interest, such as a resonant frequency, a cutoff frequency, or a center frequency. Mathematical models of nonideal inductors and capacitors which may be used for a range of frequencies about ω_0, in this case, are shown in Fig. 9.6.

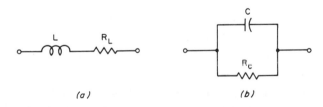

(a)   *(b)*

Figure 9.6. Nonideal models of (a) inductor, and (b) capacitor.

In the ideal case, $R_L = 0$ and R_C is infinite, in which case q_L and q_C are infinite. Practical values are $q_L = 15$ or less and $q_C = 100$ or so, depending on the quality and type.

Considering Fig. 9.6, the impedances of the nonideal elements are given by

$$Z_L(s) = sL + R_L = (s + \omega_0/q_L)L \tag{9.35}$$

and

$$Z_C(s) = \frac{1}{sC + G_c} = \frac{1}{(s + \omega_0/q_c)C} \tag{9.36}$$

We may normalize ω_0 to 1 rad/s without loss in generality, since $s + 1/q$ when denormalized (s replaced by s/ω_0) becomes $(s + \omega_0/q)/\omega_0$. Thus except for a multiplicative constant, denormalization leads to (9.35) and (9.36). In the normalized case, if we define

$$d_L = \frac{1}{q_L}, \qquad d_C = \frac{1}{q_C} \tag{9.37}$$

then (9.35) and (9.36) become

$$Z_L(s) = (s + d_L)L$$
$$Z_C(s) = \frac{1}{(s + d_C)C} \tag{9.38}$$

Hence we may account for dissipation by shifting the frequency s. The dissipation is said to be *uniform* if

$$d_L = d_C \triangleq d \tag{9.39}$$

and *nonuniform* otherwise.

The general cases of doubly-terminated and singly-terminated ladders have been considered for both uniform and nonuniform dissipation [De], [Ge-2]. However, we shall consider here only the case of uniform dissipation applied to single-termination ladders.

To obtain more practical networks using nonideal inductors and capacitors, we may use the method of *predistortion*. This procedure consists of predistorting the network function $H(s)$ by replacing s by $s - d$, realizing the resulting function $H'(s) = H(s - d)$ as an LC two-port, say, terminated in a resistor, and finally removing the effect of the predistortion by replacing s by $s + d$. This may be effected in the network by placing a resistance of Ld in series with each L and a conductance of Cd in parallel with each C, as (9.38) and Fig. 9.6 indicate. In the predistortion step, replacing s by $s - d$ shifts the poles and zeros of $H(s)$ d units to the right. Thus d cannot be so large as to shift poles into the right-half plane, nor, if a ladder is to be realized, to shift zeros into the right-half plane.

As an example, let us consider the function

$$H(s) = \frac{V_2(s)}{V_1(s)} = \frac{2}{s^2 + 2s + 2} \tag{9.40}$$

If $q_L = q_c = 4$, then $d = 1/4$, and we have $d = d_L = d_c = \frac{1}{8_L} = \frac{1}{8_c}$

$$H'(s) = H(s - 1/4) = \frac{2}{(s - 1/4)^2 + 2(s - 1/4) + 2}$$

$H'(s) = H(s - d)$

$$= \frac{2}{s^2 + \frac{3}{2}s + \frac{25}{16}}$$

Realizing this voltage ratio as an LC two-port terminated in 1Ω is accomplished by noting that

$$y_{22} = \frac{s^2 + \frac{25}{16}}{\frac{3}{2}s}$$

→ this value is the value of the transfer function of the numerator of the new transfer function (see, p 182 9-41)

with a Cauer 1 development,

$$\frac{3}{2}s) s^2 + \frac{25}{16} (\frac{2}{3}s$$

$$\frac{s^2}{\frac{25}{16}) \frac{3}{2} s (\frac{24}{25}s}$$

$$\frac{\frac{3}{2}s}{\equiv}$$

The network for $H'(s)$ is shown in Fig. 9.7.

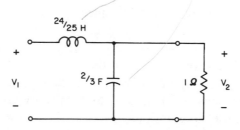

Figure 9.7. The predistorted network.

$R = L \cdot d$

Finally, to realize (9.40), we insert a resistance of $R = \dfrac{24}{25} \cdot \dfrac{1}{4} = \dfrac{6}{25}$

in series with the inductor and a conductance of $G = \dfrac{2}{3} \cdot \dfrac{1}{4} = \dfrac{1}{6}$, or

a resistance of 6 in parallel with the capacitor. The result is shown in Fig. 9.8, and analysis shows it to have a transfer function given by

$$\frac{V_2}{V_1} = \frac{25/16}{s^2 + 2s + 2} \qquad (9.41)$$

We note that the two-port obtained in Fig. 9.8 is a "lossy" twoport; that is, it contains practical inductors and capacitors. We should observe also that the gains in (9.40) and (9.41) are not the same. In (9.40) the

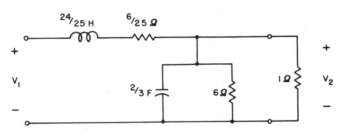

Figure 9.8. A "lossy" realization of Eq. (9.40).

gain is 1 and the loss at $s = 0$ is $\alpha = 0$ dB. In (9.41) the gain is $\frac{25}{32}$ and the loss at $s = 0$ is

$$\alpha = -20 \log \frac{25}{32} = 2.144 \text{ dB}$$

$\dfrac{V_2}{V_1}\Big|_{s=0} = \dfrac{25/16}{2} = \dfrac{25}{32}$

This is a constant loss added at all frequencies, called a *flat loss*, which is a consequence of the predistortion process. (An *LC* 2-port with a 1Ω termination realizing (9.40) would yield $H(0) = 1$ and hence $\alpha = 0$.)

We should observe at this point that the example of Fig. 9.8 is more of a learning aid than a practical illustration. To obtain Fig. 9.8 we have assumed constant q_L and q_C and uniform dissipation. These are often unrealistic assumptions, particularly in the latter case. A more realistic assumption is that q_L and q_C are approximately constant in a very narrow-band application and that $q_C \gg q_L$.

9.5 SUMMARY

In this chapter doubly-terminated networks, consisting of an LC two-port with a load resistor and a source resistor, were considered. The method discussed was the insertion-loss procedure given in general by Darlington. Also considered was the concept of predistortion, which is a means of taking into account the lossy, and thus nonideal, nature of inductors and capacitors.

This chapter concludes our discussion of passive synthesis methods. In the remainder of the book we shall consider active synthesis procedures, and in particular, the active synthesis of frequency-selective, phase-shifting, and time-delay filters.

EXERCISES

9.1. Obtain a doubly-terminated realization for the following, where R_1 and R_2 are as indicated:

$$k = \frac{2\sqrt{R_1 R_2}}{R_1 + R_2}$$

(a) $|t(j\omega)|^2 = \dfrac{k^2}{1 + \omega^4}$, $R_1 = R_2 = 1\,\Omega$

(b) $|t(j\omega)|^2 = \dfrac{k^2}{1 + \omega^4}$, $R_1 = 1\,\Omega, R_2 = 3\,\Omega$

$$f_2 = \frac{2\sqrt{R_1 R_2}}{R_1 + R_2}$$

$$|Q(j\omega)|_{min} = 1$$

Answer:

$$\frac{1}{60}$$

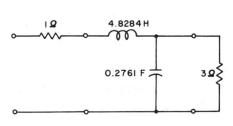

Figure Ex. 9.1.

(c) $|t(j\omega)|^2 = \dfrac{1}{1 + \omega^6}$, $R_1 = R_2 = 2\,\Omega$

(d) $|t(j\omega)|^2 = \dfrac{k^2}{1 + (\omega/2)^6}$, $R_1 = R_2 = 1\,\Omega$

(e) $|t(j\omega)|^2 = \dfrac{k^2}{1 + \omega^6}$, $R_1 = 1\,\Omega, R_2 = 3\,\Omega$

9.2. The doubly-terminated network with $R_1 = R_2 = 1\Omega$ may take on either of the two forms as shown in Fig. Ex. 9.2.

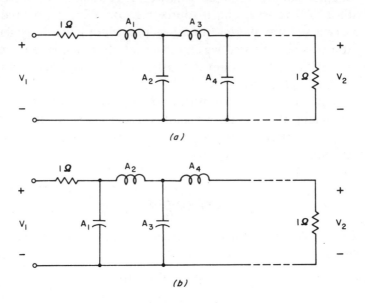

Figure Ex. 9.2.

(The transfer function is the all-pole function V_2/V_1.) It is known [Ge-1] that in the case of the normalized low-pass Butterworth filter of order n, the element values, in henrys and farads, are given by

$$A_k = 2\sin\frac{(2k-1)\pi}{6}; \qquad k = 1, 2, \ldots, n.$$

(a) Verify Fig. 2.4 for the case $n = 3$.

(b) Obtain Butterworth filters of orders 2 and 4.

9.3. For the Butterworth function

$$\frac{V_2}{V_1} = \frac{1}{s^3 + 2s^2 + 2s + 1}$$

use the method of predistortion to obtain a singly-terminated realization with a terminating resistance of 1Ω and a uniform dissipation constant of $d = \frac{1}{4}$. Calculate the flat loss.

9.4. Repeat Exercise 9.3 for a third-order Bessel filter with a gain of 1.

10

Elements
of Active Synthesis

10.1 INTRODUCTION

In the design of filters for use at low frequencies, say 1 Hz to about $\frac{1}{2}$ MHz, inductors become impractical because of their size and considerable departure from ideal behavior. Also inductors are not readily adaptable to integrated circuit techniques, which have become very important in recent years. Therefore for some time circuit designers have been developing methods of retaining the effect of inductors while avoiding their actual use. The methods come under the general heading of *active network synthesis*, in which the circuit elements used are resistors, capacitors, and one or more active devices, such as controlled sources, negative impedance converters, gyrators, operational amplifiers, etc. (A gyrator is theoretically a passive device but essentially all practical realizations of it are active.)

An example of an active synthesis problem was given earlier in Exercise 2.7. The circuit in that case was made up of resistors, capacitors, and a *voltage-controlled voltage source (VCVS)*, the general ideal case of which is shown in Fig. 10.1(a), with a standard representation shown in Fig. 10.1(b). The defining equations are

$$i_1 = 0, \qquad v_2 = \mu v_1 \qquad (10.1)$$

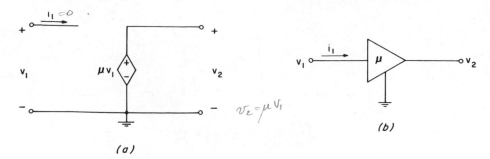

$v_2 = \mu v_1$

(b)

(a)

Figure 10.1. (a) An ideal *VCVS*, and (b) its symbolic representation.

where μ is a constant, denoted as the *gain* of the *VCVS*. In addition, the output impedance of the *VCVS* is zero, as is evident in Fig. 10.1(a).

In Fig. 10.1(b) the voltages are referred to ground, as indicated. For brevity we shall omit henceforth the connection to ground, which will be understood. A more general case may be considered, of course, where the two ports do not have a common terminal. In this case the circuit is an ideal *VCVS* with *differential* input, shown in Fig. 10.2(a). Its defining equations are

$$i_1 = i_2 = 0$$

$$v_3 = \mu(v_1 - v_2) = \mu v_d \tag{10.2}$$

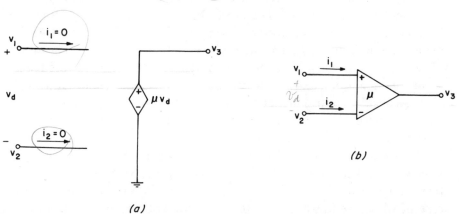

(a)

(b)

Figure 10.2. (a) An ideal differential amplifier, and (b) its symbolic representation.

the two ports do not have common terminal

(handwritten at top: Ex 2.7 circuit diagram with v_1, $3\,\Omega$, $1\,\Omega$, $1\,F$, $1F$, v_y, $2v_3$, v_2)

where v_d is the differential voltage that is being amplified. The symbol for the differential amplifier is shown in Fig. 10.2(b).

Using Fig. 10.1(b) we may redraw the circuit of Exercise 2.7 in the form of Fig. 10.3. Writing node equations in the s-domain at nodes a and b results in

$$V_a\left(\frac{1}{R_1}+\frac{1}{R_2}+\frac{1}{cs}\right)-\left(\frac{1}{R_2}\right)V_b-\left(\frac{1}{cs}\right)V_2-\left(\frac{1}{R_1}\right)V_1=0$$

(handwritten: Capacitor reciprocal at node equate $=cs$)

$$(s+4)V_a-3V_1-sV_2-V_b=0 \quad \text{①}$$

$$(s+1)V_b-V_a=0 \quad \text{②}$$

$$V_b\left(\frac{1}{R_2}+\frac{1}{c_2 s}\right)-\left(\frac{1}{R_2}\right)V_a=0$$

(handwritten left margin:)

$$V_a\left(\frac{1}{R_1}+\frac{1}{R_2}+\frac{1}{cs}\right)$$

$$=V_a\left(\frac{1}{1/3}+\frac{1}{1}+\frac{1}{\frac{1}{s}}\right)$$

$$=V_a(3+1+s)$$

(handwritten circuit figure: $C_1=1F$, $R_F=\frac{1}{3}\Omega$, $R_2=1\Omega$, V_i, V_a, V_b, amplifier $2=A$, V_2, $C_2=1F$, $V_2=2V_b$)

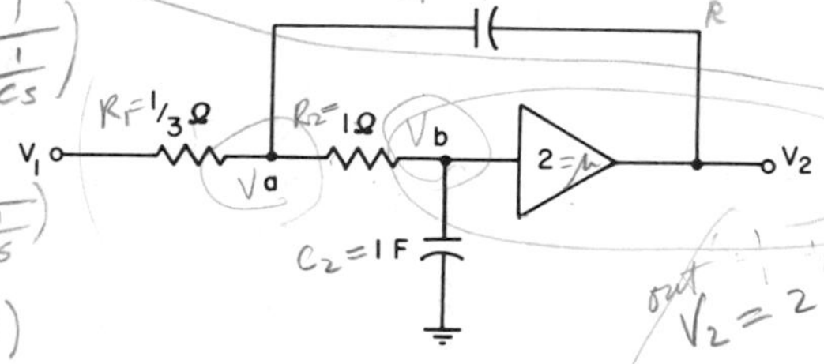

Figure 10.3. An active realization of a second-order Bessel filter.

Eliminating V_a and noting that $V_b=V_2/2$, we may write

(handwritten: eq ② 에 $(s+4)\frac{v_2}{2}$ ... $\frac{1}{2} v_4 \oplus + \oplus$ 대입)

$$\frac{V_2}{V_1}=\frac{6}{s^2+3s+3}$$

(handwritten right margin: constant 가 $\frac{1}{2}$ 이니 이거이 ; Bessel filter ; $\left.\frac{V_2}{V_1}\right|_{s=0}$)

We recognize this as the transfer function of a second-order Bessel filter (see Sec. 7.7), which if passively realized requires an RLC network. We note also that the gain realized by Fig. 10.3 is $G=2$, which is not possible with a passive realization. (By Exercise 5.9, the gain of the second-order Bessel is $G\le1$.)

There are many methods of actively synthesizing network functions, as the list of available active devices indicates. We shall restrict our attention to RC-operational amplifier circuits. That is, the active device we shall use will be the integrated circuit (IC) operational amplifier (op amp). Before proceeding to the active synthesis procedures, we shall give in the next two sections a brief discussion of the op amp and its properties. The reader who is interested in a more thorough discussion, as well as in other active devices, may consult such references as [M-1], [GTH], [S-2], [MG], [M-3], [Ha], [Hu-1], [Hu-2], and [Bu-2].

10.2 IDEAL OPERATIONAL AMPLIFIERS

Before proceeding to the operational amplifier itself, let us consider the network N of Fig. 10.4, which is a voltage amplification device with input resistance r_i, output resistance r_0, and gain constant μ. The voltage v_d is the differential voltage $v_1 - v_2$, where v_1 and v_2 are respectively the node voltages referred to ground, of nodes 1 and 2.

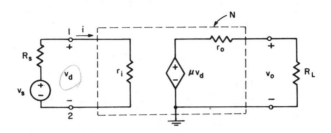

Figure 10.4. A voltage amplification network.
Ideal operational amplifier

Analyzing Fig. 10.4 yields

$$v_d = \frac{r_i}{R_s + r_i} v_s = \frac{v_s}{1 + \dfrac{R_s}{r_i}} \tag{10.3}$$

and

$$v_0 = \frac{R_L}{r_0 + R_L} \mu v_d = \frac{\mu v_d}{1 + \dfrac{r_0}{R_L}} \tag{10.4}$$

We note that if $r_i \gg R_s$ and $r_0 \ll R_L$, then $v_d \approx v_s$ and $v_0 \approx \mu v_d$, and thus N approaches an ideal *VCVS*. If node 2 is not a ground node, then the *VCVS* has differential input. In the ideal case, of course, $r_i = \infty$ (open circuit) and $r_0 = 0$ (short circuit). In the nonideal case we should select the *VCVS* or the generator and load so that the inequalities for r_i and r_0 are satisfied.

Let us consider now the network of Fig. 10.5, utilizing network N, resistor R_1, and a *feedback* resistor R_2. The input voltage is v_i and v_0 is the output voltage across the load resistor R_L. Writing node equations at nodes 1 and 3 yields

$$\left[\frac{1}{R_1} + \frac{1}{R_2} + \frac{1}{r_i} \right] v_d - \frac{1}{R_1} v_i - \frac{1}{R_2} v_0 = 0$$

feedback resistor

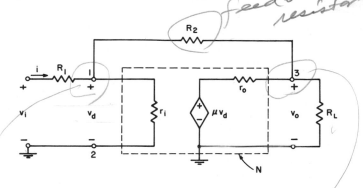

Figure 10.5. A network utilizing network N.

and

$$\left[\frac{1}{r_0} + \frac{1}{R_2} + \frac{1}{R_L}\right]v_0 - \frac{1}{R_2}v_d - \frac{\mu v_d}{r_0} = 0$$

Eliminating v_d and solving for the transfer function results in

$$\frac{v_0}{v_i} = \frac{r_0 + \mu R_2}{\left[R_1 + R_2\left(1 + \frac{R_1}{r_i}\right)\right]\left(\frac{r_0}{R_L} + 1 + \frac{r_0}{R_2}\right) - \left(\frac{R_1 r_0}{R_2} + \mu R_1\right)} \quad (10.5)$$

Suppose now that $r_i \to \infty$ and $r_0 \to 0$. In this case the network N becomes an ideal *VCVS*, as was noted earlier, v_0 becomes independent of the load R_L, and (10.5) becomes

$$\frac{v_0}{v_i} = \frac{\mu R_2}{R_1 + R_2 - \mu R_1} = \frac{R_2}{\dfrac{R_1 + R_2}{\mu} - R_1}$$

If, in addition, $\mu \to \infty$, then we have

$$\frac{v_0}{v_i} = -\frac{R_2}{R_1} \quad (10.6)$$

and thus the network of Fig. 10.5 is a voltage amplifier, though it does not have the characteristic, $i = 0$.

Under the conditions specified in the case just considered, the device N of Fig. 10.4 has the following properties: *for Ideal*

(1) It has infinite input impedance; thus the current entering either terminal 1 or terminal 2 is zero.

(2) It has zero output impedance; thus the output voltage is independent of the current drawn from terminal 3.

(3) It has infinite gain.

Since $\mu = \infty$, if $v_d = 0$, then v_0 is an indeterminate form. Therefore we shall define an additional property, known as *zero-offset*:

(4) If $v_d = 0$, then $v_0 = 0$.

Also before we pass to the limit, $\mu \rightarrow \infty$, v_0/v_d is constant (an all-pass function); thus we have already tacitly ascribed one more property, namely

(5) It has infinite bandwidth.

(*Note:* As we shall see, the first three properties are important to the circuit analyst. The last two are desirable from the standpoint of the electron device designer.)

If the network N has these five properties, then it is called an *ideal operational amplifier*, or ideal op amp, and is denoted by the symbol of Fig. 10.6. Terminal 1 is the *inverting input* terminal, terminal 2 is the *noninverting input* terminal, and terminal 3 is the *output* terminal.

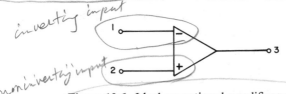

Figure 10.6. Ideal operational amplifier symbol.

A few comments are in order about the reversal of signs on the terminals 1 and 2 in Figs. 10.5 and 10.6. In actual op amp realizations the gain μ is a large negative number (at zero frequency). By retaining μ as positive and reversing the polarities of the input terminals, we have for $r_0 = 0$,

$$v_3 = \mu(v_2 - v_1) \qquad (10.7)$$

which is equivalent to Fig. 10.5 with $\mu < 0$. However in (10.7) the reader does not need to remember how the terminals are marked. A glance at Fig. 10.6 shows that v_2 is associated with plus and v_1 with minus, as (10.7) indicates.

There is a more compelling physical reason for grounding terminal 2 (the noninverting terminal) rather than terminal 1 in Fig. 10.5. The op amp is a high-gain amplifier and resistive feedback to the noninverting terminal alone (positive feedback) adds a very large signal in phase with the other incoming signal and drives the op amp to the limit of its dynamic operating range. The circuit will either oscillate or the op amp will "latch up" (its output voltage assumes the value of one of the power supplies). The reader is asked to show that the circuit is unstable for a simple example in Exercise 10.1.

Because of the infinite-gain feature of the ideal op amp, a finite output voltage v_0 corresponds to an input differential voltage of $v_d = 0$. Thus if the op amp is operated with the noninverting terminal grounded (single-input mode), then the inverting terminal is said to be at *virtual ground*. That is, it appears as if it were grounded, though it is not. In any case, in circuits containing ideal op amps, the circuit analyst may adopt as a strategem, zero currents into both input terminals and zero voltage across these terminals.

As an example, let us reconsider Fig. 10.5, redrawn using Fig. 10.6, as shown in Fig. 10.7. The node equation at node a is

$$\frac{v_i}{R_1} + \frac{v_o}{R_2} = 0$$

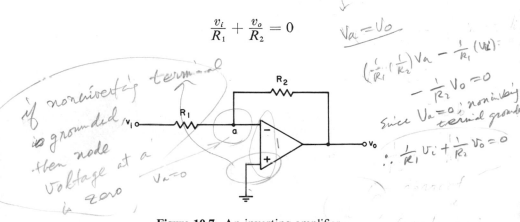

Figure 10.7. An inverting amplifier.

from which (10.6) follows. We have used the fact that the current into the op amp and the voltage with respect to ground at node a are both zero. From (10.6) we see that the gain is an *inverting* gain (negative) of magnitude R_2/R_1, and consequently Fig. 10.7 is sometimes referred to as an *inverting amplifier*, or simply an *inverter*.

Finally, we should note that (10.6) holds for Fig. 10.7 regardless of the secondary termination, open-circuited or not. This is a consequence of the zero-output impedance property of the op amp.

10.3 NONIDEAL OP AMPS

In practice, nonideal op amps may be constructed to give, for certain frequency ranges, very good approximations to ideal op amps. The gain of the op amp is, however a function of frequency and thus the op amp's behavior approaches ideal only for a specified region of operation. To illustrate this, consider the frequency domain model of a nonideal op amp shown in Fig. 10.8. The resistances $r_i = \infty$ and

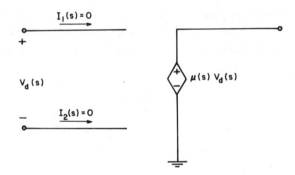

Figure 10.8. A simple model of a nonideal op amp.

$r_0 = 0$, as in the ideal case, and thus $I_1(s) = I_2(s) = 0$. However, $V_d(s)$ is not zero because the gain $\mu(s)$ is a finite function of frequency. If $\mu(s)$ is represented by

$$\mu(s) = \frac{\mu_0}{1 + \dfrac{s}{\omega_c}} \tag{10.8}$$

then Fig. 10.8 is a *one-pole rolloff* model.

We note from (10.8) that $\mu(s)$ is the transfer function of a first-order Butterworth low-pass filter. More sophisticated and more accurate models may be considered, of course, but for our purposes the first-order model will suffice. Writing (10.8) in the form

$$\mu(s) = \frac{K}{s + \omega_c} \tag{10.9}$$

we see that

$$K = \mu_0 \omega_c \qquad (10.10)$$

and is hence a *gain-bandwidth* product. A typical value of μ_0 is 10^5 and of K is 10^6; that is, ω_c is typically 10 rad/s [Bu-2].

The amplitude $|\mu(j\omega)|$ is called the *open-loop gain*, since it is the gain with no external feedback applied. Its plot is shown in Fig. 10.9.

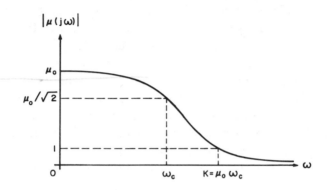

Figure 10.9. Open-loop gain of a simple op amp model.

Since the gain is a Butterworth response it has the maximally-flat characteristic near $\omega = 0$. Also we note that the open-loop gain at $\omega = K$ is given by

$$|\mu(jK)| = \frac{\mu_0}{\sqrt{\mu_0^2 + 1}} \approx 1 \qquad (10.11)$$

This is down 100 dB, if $\mu_0 = 10^5$, from its value at $\omega = 0$.

In linear operations, such as in filter theory, the op amp is never used in the open-loop mode for a variety of reasons, the foremost being its large gain property [Bu-2]. When feedback is applied the characteristics of the op amp are determined largely by the feedback elements.

Actual op amps have eight or more terminals, but we shall use the symbol of Fig. 10.6, showing only the inverting input, noninverting input, and output terminals. The other terminals not shown are two power supply terminals (± 15 V for example, to ground), *offset-null* terminals, and *compensation* terminals. Offset-null terminals are available on some op amps, such as the type 741, for the purpose of nulling the output voltage when the input terminals are grounded. Compensation terminals are for the purpose of connecting an external *compensating* network, as specified by the manufacturer, to increase the passband of the open-loop gain. Op amps like the type 709 must be externally compensated, whereas those like the type 741 are internally compensated and have no compensation terminals.

We should add that some op amps are manufactured with a single input terminal, and some have differential output terminals. Nearly all have a differential input and a single output, however.

The circuit designer should be aware of the op amp's limitations. For example, if the frequency range of operation is such that the open-loop gain is considerably below its maximum value of, say 10^5, then it may not be possible to ignore certain terms in the transfer function, as was done in going from Eq. (10.5) to Eq. (10.6). The input signal that can be applied to an op amp is also limited as shown by the manufacturer's specified *slew rate*, which is the maximum time rate of change of the output that the op amp can accomodate. For example, if $V_p \sin \omega t$ is the output voltage, then the maximum time rate of change is ωV_p V/s (occurring at $t = 0$), which for linear operation cannot exceed the slew rate of the op amp.

10.4   OP AMP APPLICATIONS

We have seen in Fig. 10.7 of Sec. 10.2 how an op amp may be used with two resistors to obtain an inverting amplifier. A *noninverting* amplifier may also be obtained from an ideal op amp and two resistors, as shown in Fig. 10.10. The node equation at node a is

$$\left(\frac{1}{R_1} + \frac{1}{R_2}\right) V_1 - \frac{1}{R_2} V_2 = 0 \tag{10.12}$$

since the current into the noninverting terminal is zero and the voltage at node a is equal to V_1 because of the zero voltage feature between the input terminals. From (10.12) we have

$$V_2 = \mu V_1 \tag{10.13}$$

where

$$\mu = 1 + \frac{R_2}{R_1} \tag{10.14}$$

Thus since $I = 0$, Fig. 10.10 is an ideal *VCVS* with gain $\mu \geq 1$.

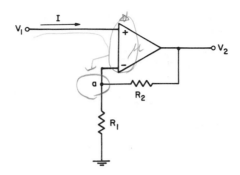

Figure 10.10. A noninverting amplifier, or *VCVS*.

Unity gain, $\mu = 1$, is achieved in Fig. 10.10 if $R_1 = \infty$ (open-circuit) or $R_2 = 0$. This may be achieved by connecting the inverting terminal directly to the output terminal, as shown in Fig. 10.11. Such a device is called a *voltage follower*, since the output voltage follows the input voltage exactly. It may also be used as a *buffer amplifier* to provide

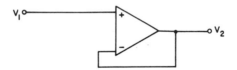

Figure 10.11. A voltage follower.

isolation of source and load, thus preventing loading effects. For example, if a circuit is designed for an output voltage across a pair of open-circuited terminals, but the load to be connected at the output port does not have infinite input impedance, then a buffer amplifier may be placed between the designed circuit and the load.

Let us next consider the circuit of Fig. 10.12, which has n voltage sources connected through resistors to the inverting input terminal. The node equation at this terminal is

$$-\left(\frac{V_1}{R_1} + \frac{V_2}{R_2} + \cdots + \frac{V_n}{R_n}\right) - \frac{V_0}{R_0} = 0$$

from which we have

$$V_0 = -R_0 \sum_{i=1}^{n} \frac{V_i}{R_i} \qquad (10.15)$$

Thus Fig. 10.12 is a *summing* amplifier.

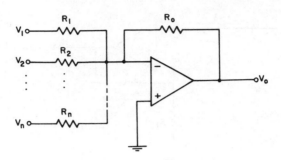

Figure 10.12. A summing amplifier.

As a final example, let us consider the circuit of Fig. 10.13. At the noninverting terminal we have

$$-\frac{V_1}{R} - sCV_2 = 0$$

or

$$V_2 = -\frac{1}{RCs}V_1 \tag{10.16}$$

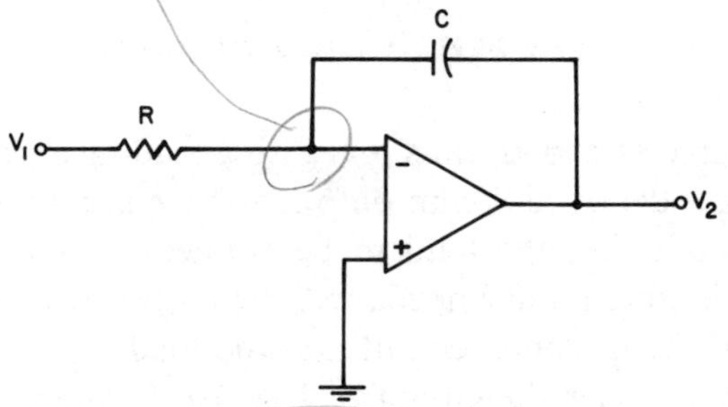

Figure 10.13. An integrator.

Thus Fig. 10.13 is an *integrator*, since in the time domain, (10.16) becomes

$$v_2(t) = -\frac{1}{RC} \int_0^t v_1(t)\, dt$$

(assuming zero voltage on the capacitor at $t = 0$).

10.5 INFINITE-GAIN, MULTIPLE-FEEDBACK FILTERS

this chapter

In this section we shall consider a general circuit of Rs, Cs, and one op amp, which can be used to realize second-order filter transfer functions of various types [GTH]. The circuit is shown in Fig. 10.14, where the admittances are either $Y_i = G_i = 1/R_i$ or $Y_i = C_i s$, which, properly chosen, yield low-pass, high-pass, or bandpass filters of order two. The circuit is called an *infinite-gain, multiple-feedback* (*MFB*) network because of the multiple feedback paths and the fact that the op amp is operating in an infinite gain rather than a finite gain (*VCVS*) mode.

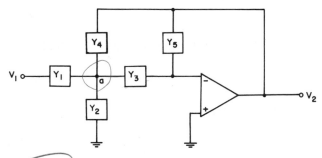

Figure 10.14. A general second-order infinite-gain *MFB* filter.

To see that Fig. 10.14 is a general filter circuit, let us write the node equations at the intermediate node a and the inverting input node, resulting in

$$Y = \frac{1}{Z} = \frac{1}{R_i} = C_i s$$

$$(Y_1 + Y_2 + Y_3 + Y_4)V_a - Y_1 V_1 - Y_4 V_2 = 0$$

and

$$- Y_3 V_a - Y_5 V_2 = 0$$

Eliminating V_a we may write (by simultaneous equation)

$$\frac{V_2}{V_1} = -\frac{Y_1 Y_3}{Y_5(Y_1 + Y_2 + Y_3 + Y_4) + Y_3 Y_4} \qquad (10.17)$$

We note that in all cases the circuit produces an inverting gain.

$$\frac{\frac{1}{R_1}\frac{1}{R_3}}{C_5 s\left(\frac{1}{R_1}+C_2 s+\frac{1}{R_3}+\frac{1}{R_4}\right)+\frac{1}{R_3}\frac{1}{R_4}}$$

Low pass function

$$H(s) = \frac{Gb_0}{s^n + b_{n-1}s^{n-1} + \cdots + b_1 s + b_0}$$

for low pass filter

If a low-pass second-order filter is desired we need, from (4.7) for an inverting gain, the transfer function

$b_0 = $ constant   *p?l*

$G = Y = \frac{1}{Rc}$

$$\frac{V_2}{V_1} = -\frac{Gb_0}{s^2 + b_1 s + b_0} \qquad (10.18)$$

compare w/ 10.17, numerator

$Y_1 \& Y_3$ are $\frac{1}{Rc}$

where G is the gain and the b_0 and b_1 determine the type of low-pass filter. Comparing (10.17) and (10.18) we see that we must have

b_0 is constant

$$Y_5 = C_5 s \qquad \qquad (10.19)$$

this gives

$Y_3 \, Y_4$ are $\frac{1}{Rc}$

and that Y_1, Y_3, and Y_4 must be constants; that is

$$Y_1 = G_1, \; Y_3 = G_3, \; Y_4 = G_4 \qquad (10.20)$$

Since, this is 2nd order

Finally, to achieve the quadratic denominator it is necessary to have

require 2 poles

$= 2$ capa.

$$Y_2 = C_2 s \qquad (10.21)$$

$C_5 \& C_2$

The transfer function is then

$= 1$ cap

$$\frac{V_2}{V_1} = -\frac{\dfrac{G_1 G_3}{C_2 C_5}}{s^2 + \left(\dfrac{G_1 + G_3 + G_4}{C_2}\right)s + \dfrac{G_3 G_4}{C_2 C_5}}$$

Renumbering the elements as shown in Fig. 10.15 yields the transfer function (10.18), where

$$b_0 = \frac{1}{R_2 R_3 C_1 C_2}$$

$$b_1 = \frac{1}{C_2}\left(\frac{1}{R_1} + \frac{1}{R_2} + \frac{1}{R_3}\right) \qquad (10.22)$$

$$G = \frac{R_2}{R_1}$$

fw High pass filter

MFB

If we wish Fig. 10.14 to be a high-pass filter then we must have (10.17) reduce to (4.8) for $n = 2$, given for inverting gain by

for High pass Y_1, Y_3, Y_5 order

$$\frac{V_2}{V_1} = -\frac{Gs^2}{s^2 + a_1 s + a_0} \qquad (10.23)$$

Gs^2 requires $Y_1 \& Y_5$ are capacitors

$Y_1 = G_1 = C_1 s$ $Y_3 = G_3 = C_3 s$

$a_0 = Y_3 Y_4 \Rightarrow$ gives $Y_4 = $ capacitors

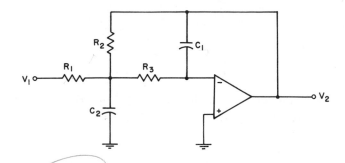

Figure 10.15. An infinite-gain *MFB* low-pass filter.

Thus both Y_1 and Y_3 must represent capacitors and Y_5 must represent a resistor. These require that Y_4 represent a capacitor and Y_2 a resistor. The circuit, with the elements numbered as shown, is given in Fig. 10.16, and (10.23) holds for

$$a_0 = \frac{1}{R_1 R_2 C_2 C_3}$$

$$a_1 = \frac{C_1 + C_2 + C_3}{R_2 C_2 C_3} \tag{10.24}$$

$$G = \frac{C_1}{C_2}$$

We may note that the high-pass filter of Fig. 10.16 may also be obtained by applying the $RC : CR$ transformation of Sec. 4.6 to the low-pass filter of Fig. 10.15.

Finally, to obtain a second-order bandpass filter with inverting gain

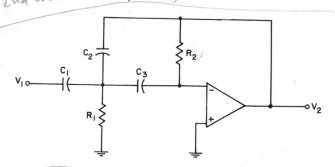

Figure 10.16. An infinite-gain *MFB* high-pass filter.

for 2nd order Bandpass filter

G, center frequency ω_0, and bandwidth B, by (4.7) and (4.15) we must have

$B = \dfrac{\omega_0}{Q}$

$$\frac{V_2}{V_1} = \frac{-GBs}{s^2 + Bs + \omega_0^2} \tag{10.25}$$

There are several ways of achieving this with Fig. 10.14, one of which is shown in Fig. 10.17, where (10.25) is realized with

p74

$$B = \frac{C_1 + C_2}{R_3 C_1 C_2}$$

$$\omega_0^2 = \frac{1}{R_3 C_1 C_2}\left(\frac{1}{R_1} + \frac{1}{R_2}\right) \tag{10.26}$$

$$G = \frac{R_3 C_2}{R_1(C_1 + C_2)}$$

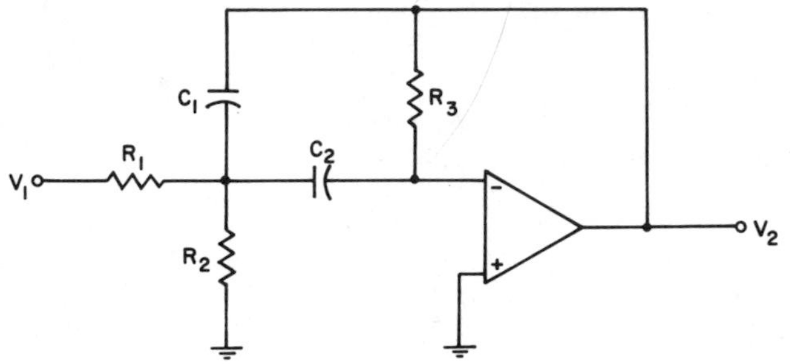

Figure 10.17. An infinite-gain *MFB* bandpass filter.

10.6 GENERAL *VCVS* FILTERS

Another general filter structure which may be used to obtain second-order low-pass, high-pass, and bandpass active filters is the network of Fig. 10.18. We shall call this a *VCVS* filter because of the presence of the ideal *VCVS* with gain $\mu > 0$. The better known filters of this type were first obtained by Sallen and Key [SK] in 1955. We shall consider only the low- and high-pass filters here, and give a modification later in Sec. 11.11, in which there is a Y_s to ground from node *a*, realizing a *VCVS* bandpass filter.

The node equations at nodes *a* and *b* are given by

at node a

$$(Y_1 + Y_2 + Y_3)V_a - Y_1 V_1 - \frac{Y_2 V_2}{\mu} - Y_3 V_2 = 0$$

$-Y_2 V_b$ *where* $V_2 = \mu V_b$

$\therefore V_b = \dfrac{V_2}{\mu}$

[handwritten top:] From low pass filter ; 2nd order
$$H(S)=\frac{V_2}{V_1}=+\frac{G b_0}{s^2+b_1 s+b_0} \quad \text{VCVS}$$
For Hi pass filter ; 2nd order
$$H(S)=\frac{V_2}{V_1}=-\frac{G s^2}{s^2+a_1 s+a_0}$$

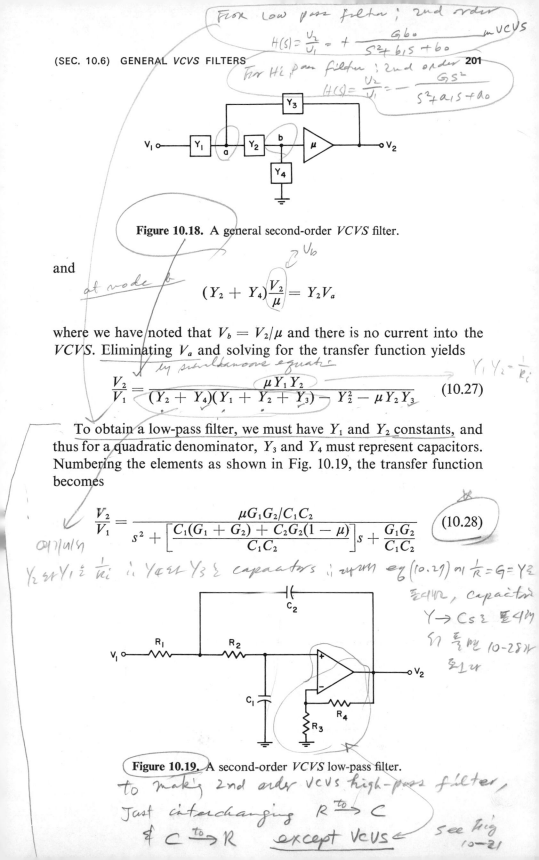

Figure 10.18. A general second-order *VCVS* filter.

and

[handwritten:] at node b
$$(Y_2 + Y_4)\frac{V_2}{\mu} = Y_2 V_a$$

where we have noted that $V_b = V_2/\mu$ and there is no current into the *VCVS*. Eliminating V_a and solving for the transfer function yields

[handwritten:] by simultaneous equation $Y_1 Y_2 = \frac{1}{R_i}$

$$\frac{V_2}{V_1} = \frac{\mu Y_1 Y_2}{(Y_2 + Y_4)(Y_1 + Y_2 + Y_3) - Y_2^2 - \mu Y_2 Y_3} \qquad (10.27)$$

To obtain a low-pass filter, we must have Y_1 and Y_2 constants, and thus for a quadratic denominator, Y_3 and Y_4 must represent capacitors. Numbering the elements as shown in Fig. 10.19, the transfer function becomes

$$\frac{V_2}{V_1} = \frac{\mu G_1 G_2/C_1 C_2}{s^2 + \left[\dfrac{C_1(G_1 + G_2) + C_2 G_2(1 - \mu)}{C_1 C_2}\right]s + \dfrac{G_1 G_2}{C_1 C_2}} \qquad (10.28)$$

[handwritten left margin:] 01/04/04

[handwritten:] Y_2 & Y_1 & $\frac{1}{R_i}$; Y_4 & Y_3 & capacitors ; 대입 eq (10.27) 에 $\frac{1}{R}=G=Y$ & 대입, capacitor $Y \to Cs$ & 대입 10-28식 & 대입

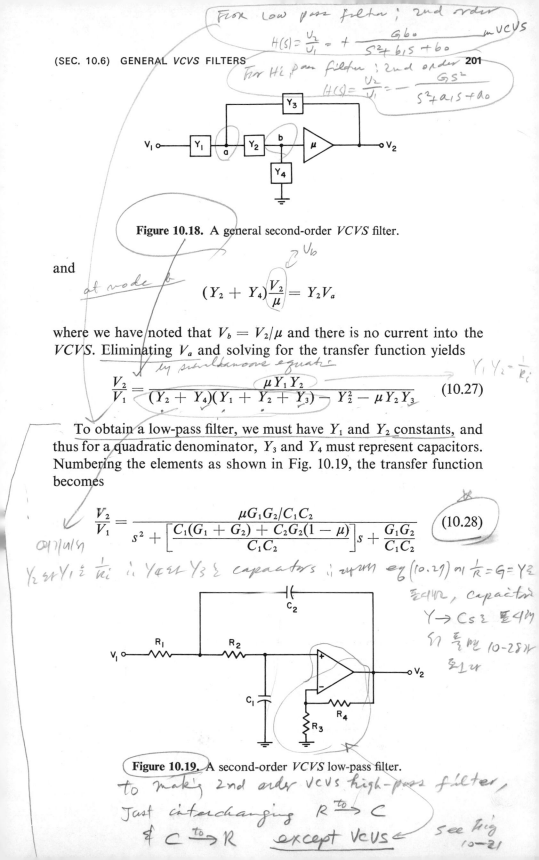

Figure 10.19. A second-order *VCVS* low-pass filter.

[handwritten bottom:] to making 2nd order VCVS high-pass filter, Just interchanging $R \xrightarrow{to} C$ & $C \xrightarrow{to} R$ except VCVS ← see fig 10-21

if ω_c is unknown, f_c is given
then $C_1 = \frac{1}{2\pi f_c k_i}$ or/if k_i ...

if ω_c is given, then $C_1 = \frac{c_i'}{\omega_c k_i}$

We have replaced the *VCVS* by its op amp realization, so that by (10.14),

$$\mu = 1 + \frac{R_4}{R_3} \tag{10.29}$$

Matching (10.28) with the general case,

$$\frac{V_2}{V_1} = \frac{G b_0}{s^2 + b_1 s + b_0} \tag{10.30}$$

we have

$$b_0 = \frac{1}{R_1 R_2 C_1 C_2}$$

$$b_1 = \frac{1}{C_2}\left(\frac{1}{R_1} + \frac{1}{R_2}\right) + \frac{1}{R_2 C_1}(1 - \mu) \tag{10.31}$$

$$G = \mu = 1 + \frac{R_4}{R_3}$$

Resistors R_3 and R_4 are arbitrary except that their ratio must be fixed. However, for best operation they should be chosen to minimize the *dc* offset of the op amp. To see how this is done, consider the circuit of Fig. 10.19 operating at *dc* (the power supply voltages) with the input and output terminals grounded. This is shown with the capacitors represented by open circuits in Fig. 10.20. Theoretically there is no voltage $V_a - V_b$, but in order for a real op amp to work there must be a slight current in its input leads. Thus the offset voltage $V_a - V_b$ will be minimized if the resistances $R_1 + R_2$ and $R_3 R_4/(R_3 + R_4)$ of the two

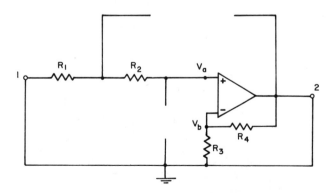

Figure 10.20. The *VCVS* filter operating at dc.

sally $C_{new} = \dfrac{C_{old}}{W_c \, k_i}$

where $W_c = 2\pi f_c$

parallel input paths are equal. From this condition and (10.29) we may write

$$\mu = 1 + \frac{R_4}{R_3} = \frac{(R_3 + R_4)R_4}{R_3 R_4} = \frac{R_4}{R_1 + R_2}$$

Thus we have for minimum *dc* offset

$$R_4 = \mu R_{eq}$$

$$R_3 = \frac{R_4}{\mu - 1} = \frac{\mu R_{eq}}{\mu - 1}, \qquad \mu \neq 1 \tag{10.32}$$

where

$$R_{eq} = R_1 + R_2 \tag{10.33}$$

Fig. 10.20 also illustrates why we cannot capacitively couple a source into node 1, for there would be no *dc* return to ground from the noninverting terminal of the op amp. Thus no current could flow and the op amp would not work. In general, there must always be a *dc* return to ground at each input terminal of the op amp.

An example of a *VCVS* low-pass filter was given earlier in Fig. 10.3, for the case $R_1 = \frac{4}{3}\Omega$, $R_2 = 1\Omega$, $C_1 = C_2 = 1F$, and $\mu = 2$. This resulted in a second-order Bessel filter. Using (10.32) and (10.33) we should have $R_3 = R_4 = \frac{8}{3}\Omega$. $\quad R_4 = \mu R_{eq} = 2(\frac{1}{3}+1)=\frac{8}{3}$, $R_3 = \frac{R_4}{\mu-1}$

$= \frac{\frac{8}{3}}{2-1}$

$= \frac{8}{3}$

By a dual development we may obtain a second-order *VCVS* high-pass filter. However, by the *RC* : *CR* transformation we know that a high-pass filter may be obtained from the low-pass filter of Fig. 10.19 by leaving the *VCVS* unchanged and interchanging resistors with capacitors and vice-versa. This results in the high-pass filter of Fig. 10.21,

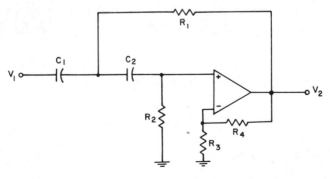

Figure 10.21. A second-order *VCVS* high-pass filter.

for which analysis yields

$$\frac{V_2}{V_1} = \frac{\mu s^2}{s^2 + \left[\dfrac{G_2(C_1 + C_2) + C_2 G_1(1 - \mu)}{C_1 C_2}\right]s + \dfrac{G_1 G_2}{C_1 C_2}} \qquad (10.34)$$

Matching (10.34) with the general case,

$$G_1 = \frac{1}{R_1},\ G_2 = \frac{1}{R_2}$$

$$\frac{V_2}{V_1} = \frac{G s^2}{s^2 + a_1 s + a_0} \qquad (10.35)$$

results in

$$a_0 = \frac{1}{R_1 R_2 C_1 C_2}$$

$$a_1 = \frac{1}{R_2}\left(\frac{1}{C_1} + \frac{1}{C_2}\right) + \frac{1}{R_1 C_1}(1 - \mu) \qquad (10.36)$$

$$G = \mu = 1 + \frac{R_4}{R_3}$$

The *dc* return to ground is satisfied by R_2, and for minimum offset R_3 and R_4 are given as before by (10.32), where

$$R_{eq} = R_2 \qquad (10.37)$$

10.7 BIQUAD FILTERS

In this section we consider a general second-order filter circuit which requires more elements than either the infinite-gain *MFB* or the *VCVS* filters of the previous two sections, but which has excellent stability and tuning features, and in the case of bandpass filters, is capable of attaining much higher values of Q. The circuit is referred to in the literature as a *biquad* circuit because versions of it can be used to realize the general biquadratic (ratio of two quadratics) transfer function [KHN], [T-1], [T-2].

Let us begin by considering the second-order high-pass function with inverting gain, which we write in the form

$$\frac{V_2}{V_1} = -\frac{K s^2}{s^2 + as + b} \qquad (10.38)$$

This may be rewritten as

$$\frac{V_2}{V_1} = -\frac{K}{1 + \dfrac{a}{s} + \dfrac{b}{s^2}} \cdot \frac{T/K}{T/K}$$

where $T = -V_2$

which is equivalent to

$$\boxed{V_2 = -T}$$

$$T = KV_1 - a\frac{T}{s} - b\frac{T}{s^2} \tag{10.39}$$

Thus (10.38) is realized if we can obtain a network which realizes the second of (10.39) with V_2 measured at the node having voltage $-T$.

The second of (10.39) is satisfied, as was shown in (10.15), by the summing amplifier of Fig. 10.22. To obtain the voltage $-T/s$, we may

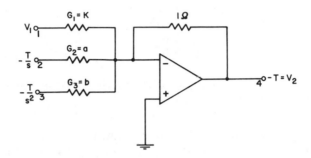

Figure 10.22. A summing amplifier which realizes Eq. (10.39).

connect an integrator with $RC = 1$ (Fig. 10.13) from node 4 to a node 5, at which the voltage is T/s. Then an inverter with gain -1 (Fig. 10.7) from node 5 to node 2 provides the voltage $-T/s$. Finally, to obtain the voltage $-T/s^2$ at node 3 we need an integrator with $RC = 1$ between nodes 3 and 5. The result is shown in Fig. 10.23, where the additional resistors and capacitors have been given the value 1. In the normalized case this is a reasonable value.

The circuit is easy to analyze since by inspection we have

$$V_3 = -\frac{V_2}{s}$$

$$V_4 = -\frac{V_3}{s} = \frac{V_2}{s^2} \tag{10.40}$$

$$V_5 = -V_3 = \frac{V_2}{s}$$

Thus writing a node equation at node 6 we have

$$-KV_1 - V_2 - \frac{aV_2}{s} - \frac{bV_2}{s^2} = 0$$

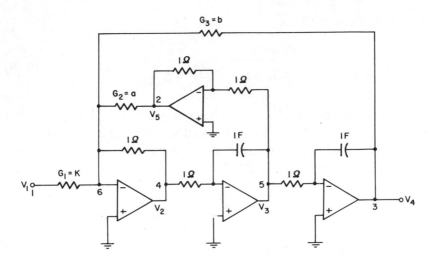

Figure 10.23. A circuit yielding low-pass, high-pass, and band-pass characteristics.

from which (10.38) follows. Thus if the output is taken at node 4, we have a high-pass filter. Similarly, in view of (10.38) and (10.40), if the output is V_3, then we have the bandpass function

$$\frac{V_3}{V_1} = \frac{Ks}{s^2 + as + b} \tag{10.41}$$

and if the output is V_4 we have the low-pass function

$$\frac{V_4}{V_1} = -\frac{K}{s^2 + as + b} \tag{10.42}$$

The steps outlined in obtaining the filter of Fig. 10.23 are equivalent to the *state-variable* method considered in reference [KHN], and may be systematically extended to other filter types as well as to higher order filters.

If we are only interested in bandpass and low-pass structures and thus have no need for node 4, we may eliminate one op amp in Fig. 10.23 by combining the operation of summing with that of integrating. This may be seen by considering the summing integrator of Fig. 10.24, for which we have

$$-KV_1 + a\frac{T}{s} + b\frac{T}{s^2} = sV_0 \tag{10.43}$$

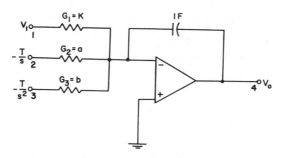

Figure 10.24. A summing integrator.

Comparing (10.43) with the second of (10.39) we see that $V_0 = -T/s$. Thus we may tie node 2 of Fig. 10.24 directly to node 4. To obtain $-T/s^2$ at node 3 we need an integrator with $RC = 1$ between node 4 and a node 5, and then an inverter with gain -1 between nodes 5 and 3. The result is shown in Fig. 10.25, where it may be seen that

$$\frac{V_2}{V_1} = -\frac{Ks}{s^2 + as + b} \tag{10.44}$$

which is a bandpass response, and

$$\frac{V_3}{V_1} = \frac{K}{s^2 + as + b} \tag{10.45}$$

which is a low-pass response.

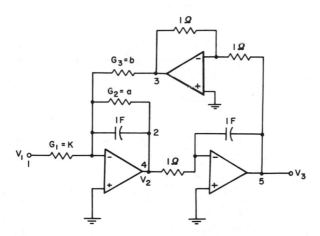

Figure 10.25. A biquad circuit yielding low-pass and band-pass characteristics.

Figs. 10.23 and 10.25 are examples of biquad circuits, Fig. 10.25 being very similar to the biquad of Tow [T-3]. In both cases we see that the given parameters K, a, and b may be obtained by adjusting respectively the conductances G_1, G_2, and G_3.

10.8 AN ALL-PURPOSE BIQUAD

In this section we shall consider the biquad circuit [FT] of Fig. 10.26, which is capable of realizing $\pm H(s)$, where

$$H(s) = \frac{V_0(s)}{V_1(s)} = -\frac{cs^2 + ds + e}{s^2 + as + b} \tag{10.46}$$

for a, b, c, d, and e real constants, and $a, b > 0$, with the single exception $ce < 0$. (This exception will be considered in Exercise 10.21.) The output V_0 may be either V_2 or V_3, depending on the numerator coefficients.

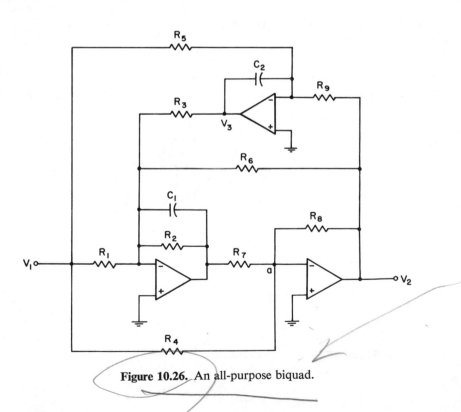

Figure 10.26. An all-purpose biquad.

Analysis of Fig. 10.26 yields

$$\frac{V_2}{V_1} = -\frac{cs^2 + ds + e}{s^2 + as + b} \qquad (10.47)$$

and

$$\frac{V_3}{V_1} = \frac{d's + e'}{s^2 + as + b} \qquad (10.48)$$

where

$$a = \frac{G_2G_8 - G_6G_7}{C_1G_8} \qquad (10.49)$$

$$b = \frac{G_3G_7G_9}{C_1C_2G_8} \qquad (10.50)$$

$$c = \frac{G_4}{G_8} \qquad (10.51)$$

$$d = \frac{G_2G_4 - G_1G_7}{C_1G_8} \qquad (10.52)$$

$$e = \frac{G_3G_5G_7}{C_1C_2G_8} \qquad (10.53)$$

$$d' = \frac{G_4G_9 - G_5G_8}{C_2G_8} \qquad (10.54)$$

$$e' = \frac{G_9(G_2G_4 - G_1G_7) - G_5(G_2G_8 - G_6G_7)}{C_1C_2G_8} \qquad (10.55)$$

For the normalized case, let us choose

$$C_1 = C_2 = G_7 = G_8 = G_9 = 1 \qquad (10.56)$$

Then (10.47) holds if from (10.49) through (10.53) we have

$$\begin{aligned}
G_1 &= c(a + G_6) - d \\
G_2 &= a + G_6 \\
G_3 &= b \\
G_4 &= c \\
G_5 &= e/b
\end{aligned} \qquad (10.57)$$

These are realizable conductances if $e \geq 0$ and $c > 0$, since G_6, which is arbitrary, may be chosen so that $G_1 \geq 0$; that is,

$$G_6 \geq \frac{d - ac}{c} \qquad (10.58)$$

If $d - ac$ is negative or zero, we may choose $G_6 = 0$, which eliminates the resistor R_6.

If $e < 0$ and $c < 0$, then we may choose (10.57) except that $G_4 = -c$, $G_5 = -e/b$, and $G_1 = -c(a + G_6) + d$, in which case (10.58) becomes

$$G_6 \geq \frac{ac - d}{c} \qquad (10.59)$$

In this case (10.47) is realized with the sign changed.

If $ce < 0$, that is, $c, e \neq 0$ and have opposite signs, then a modification of Fig. 10.26 is necessary [FT]. As stated earlier, this modification is considered in Exercise 10.21.

Let us now consider the case $c = 0$, which requires

$$G_4 = 0 \qquad (10.60)$$

Choosing (10.56) again we have from (10.57),

$$\begin{aligned}
G_1 &= -d \\
G_2 &= a \\
G_3 &= b \\
G_5 &= e/b
\end{aligned} \qquad (10.61)$$

We have taken $G_6 = 0$, since it is no longer needed to make G_1 realizable. The conductances are realizable if $d \leq 0$ and $e \geq 0$. If $d \geq 0$ and $e \leq 0$, we may take $G_1 = d$ and $G_5 = -e/b$, resulting in the negative of (10.47).

To obtain $c = 0$ and $de \geq 0$ (both with the same sign), we use (10.48), (10.49), (10.50), (10.54), and (10.55). Choosing (10.56) again we have, if $G_6 = 0$,

$$\begin{aligned}
a &= G_2 \\
b &= G_3 \\
d' &= G_4 - G_5 \\
e' &= G_2 G_4 - G_1 - G_2 G_5
\end{aligned} \qquad (10.62)$$

The latter two equations may be written

$$G_4 = d' + G_5$$
$$G_1 = ad' - e'$$

(10.63)

If $d', e' \geq 0$ and $ad' \geq e'$, these are realizable since G_5 may be selected so that $G_4 \geq 0$. If $d', e' \geq 0$ and $ad' \leq e'$, replace e' and d' by their negatives, obtaining the negative of (10.48). Evidently, if $d', e' \leq 0$, we may replace e' and d' by their negatives and repeat the foregoing procedure.

Finally we note that Fig. 10.26 may be used to obtain a low-pass filter $(c = d = 0)$, a bandpass filter $(c = e = 0)$, a high-pass filter $(d = e = 0)$, a band-reject filter $(d = 0, e = bc)$, and an all-pass filter $(d = -ac, e = bc)$. In Chapter 11, as well as in Exercises 10.23 and 10.24, we shall consider these special cases.

10.9 HIGHER-ORDER FILTERS

A higher-order transfer function $H(s)$ is one whose numerator and/or denominator polynomials are of degree greater than two. We shall consider in this section the case of a denominator with degree $n > 2$ and a numerator with degree $m \leq n$.

There are two general methods of synthesizing higher-order functions. One way is to attempt to find a single structure which realizes the function, such as a state-variable realization [KHN], or some type of multiple-feedback circuit. The first case is a generalization of the idea used to obtain the biquad circuits of Sec. 10.7, and the second is a generalization of the methods of Secs. 10.5 and 10.6 where the general structure to be matched to the function is much more complicated.

The second general method is to factor the transfer function $H(s)$ into second-order functions (if n is odd, a first-order factor will also be required), realize each factor H_i with a subnetwork, or section, or *resonator* N_i, and cascade the resonators to obtain the overall network [T-3]. The cascading procedure must leave the H_i intact so that overall we have

$$H(s) = \prod_i H_i(s)$$

(10.64)

This will be the case, of course, if the output of each N_i is the output of an op amp.

The cascaded network has the advantage that the mathematics is relatively simple, requiring only the factoring procedure and the use of familiar subnetworks with which to match the coefficients. Another advantage is that each resonator can be tuned separately to the required function H_i.

The primary disadvantage of the cascaded network, and one which has led researchers in recent years to seek alternate configurations [Sz], is its *sensitivity* properties. We shall consider sensitivity in some detail in Chapter 12, where we shall show that the overall transfer function sensitivity with respect to H_i in (10.64) is 1. That is, a relative change in H_i yields exactly the same relative change in H.

Since in the odd-order case we need to realize first-order functions in the cascade method, we shall consider some typical examples. As analysis will verify, Fig. 10.27(a) has transfer function

$$\frac{V_2}{V_1} = \frac{\mu G/C}{s + G/C} \tag{10.65}$$

and Fig. 10.27(b) has transfer function

$$\frac{V_2}{V_1} = -\frac{G_1/C}{s + G_2/C} \tag{10.66}$$

Thus these are first-order low-pass sections with noninverting and inverting gains, respectively.

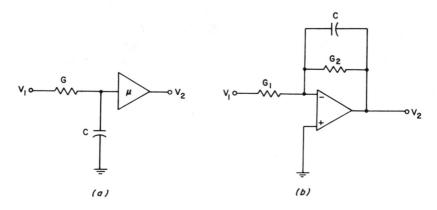

(a)   (b)

Figure 10.27. Two first-order low-pass filters.

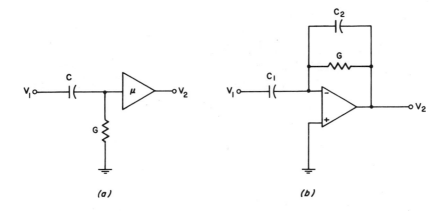

Figure 10.28. Two first-order high-pass sections.

Analysis of Figs. 10.28(a) and (b) shows that they are first-order high-pass sections having noninverting gain and inverting gain, respectively. In (a) the transfer function is

$$\frac{V_2}{V_1} = \frac{\mu s}{s + G/C} \qquad (10.67)$$

and in (b) the transfer function is

$$\frac{V_2}{V_1} = -\frac{C_1 s/C_2}{s + G/C_2} \qquad (10.68)$$

A general first-order function

$$\frac{V_2}{V_1} = -\frac{cs + d}{s + a} \qquad (10.69)$$

may be obtained from Fig. 10.26 by making $b = e = 0$ in (10.46). Considering $a, c > 0$, we have from (10.57)

$$G_1 = c(a + G_6) - d$$
$$G_2 = a + G_6$$
$$G_4 = c \qquad\qquad (10.70)$$
$$G_3 = G_5 = 0$$

(We see from (10.50) and (10.53) that $G_3 = 0$ will suffice. We also choose $G_5 = 0$ since it plays no role in the other equations.) The normalized values (10.56) and (10.70) are now used to construct the circuit from Fig. 10.26. The result is shown in Fig. 10.29, where it may be noted that $G_3 = G_5 = 0$ disconnects the integrator, which is therefore not shown.

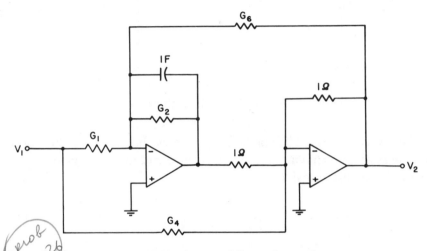

Figure 10.29. A general first-order section.

Thus given $a, c > 0$, we may select G_6, which is arbitrary, so that $G_1 > 0$. This requires that (10.58) holds. If, as before, $d - ac \leq 0$ we may take $G_6 = 0$, eliminating resistor G_6. The other elements in (10.70) are realizable.

As an example, suppose we wish to synthesize a third-order Butterworth filter with transfer function

$$\frac{V_2}{V_1} = \frac{4}{s^3 + 2s^2 + 2s + 1} \tag{10.71}$$

which we factor in the form

$$\frac{V_2}{V_1} = H_1 H_2$$

where

$$H_1 = \frac{2}{s + 1} \tag{10.72}$$

and

$$H_2 = \frac{2}{s^2 + s + 1} \tag{10.73}$$

We may realize H_1 by Fig. 10.27(a) with $G = C = 1$ and $\mu = 2$, and H_2 by Fig. 10.19 with $C_1 = C_2 = G_1 = G_2 = 1$ and $\mu = 2$. The result is shown in Fig. 10.30.

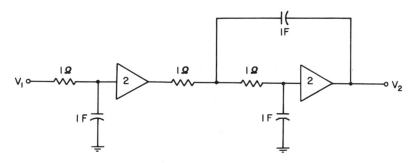

Figure 10.30. A third-order low-pass Butterworth filter.

For the reader's convenience the second-order denominator factors are given in Appendix B for the more commonly-used even-order cases of Butterworth, Chebyshev, and Bessel filters. Because it is only slightly more complicated to construct a second-order section than a first-order section, we have omitted the odd orders.

We shall consider later some examples of multiple-feedback synthesis, but to illustrate the complexity of the mathematics involved, let us consider the configuration of Fig. 10.31. This circuit is similar to

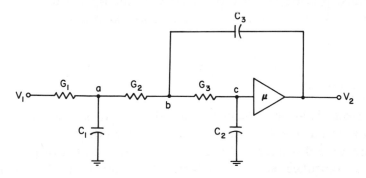

Figure 10.31. A third-order low-pass filter.

that of Fig. 10.30 except that the first *VCVS* has been removed. Since in this case the first section is loaded by the second, the overall transfer function is no longer the product of the two subnetwork functions. The figure is still a third-order low-pass filter, which may be matched to, say (10.71) by solving for the resistances and capacitances.

Writing node equations at points a, b, and c respectively, we have

$$(sC_1 + G_1 + G_2)V_a - G_1V_1 - G_2V_b = 0$$
$$(sC_3 + G_2 + G_3)V_b - G_2V_a - sC_3V_2 - G_3V_2/\mu = 0$$
$$(sC_2 + G_3)V_2/\mu - G_3V_b = 0$$

Eliminating V_a and V_b and solving for the transfer function, we have

$$\frac{V_2}{V_1} = \frac{\mu G_1 G_2 G_3}{D(s)}$$

where

$$D(s) = (sC_3 + G_2 + G_3)(sC_2 + G_3)(sC_1 + G_1 + G_2)$$
$$- G_2^2(sC_2 + G_3) - G_3(sC_1 + G_1 + G_2)(sC_3\mu + G_3)$$

Evidently the transfer function is that of a third-order low-pass filter, but to match the coefficients of the various powers of s in $D(s)$ with those of a given case like (10.71) requires the solution of a set of non-linear equations in μ, the capacitances, and the conductances. Fig. 10.31 is not, strictly speaking, a multiple-feedback circuit, because there is only one feedback path. However, it illustrates the type of mathematics involved in many cases of multiple-feedback synthesis.

As examples of multiple-feedback responses, Figs. 3.4, 3.7(b), and 4.3, shown previously, were obtained with multiple-feedback circuits of sixth and seventh orders, developed as extensions of Fig. 10.31 [HJE].

10.10 SUMMARY

This chapter is an introduction to active network synthesis, in which the active element is the operational amplifier. Both ideal and practical op amps were discussed and used to obtain various network elements such as controlled sources, integrators, summers, etc. Infinite-gain, multiple-feedback structures, *VCVS* circuits, and biquad circuits were considered in general and in particular cases. Finally, an all-purpose biquad was given and methods were discussed for obtaining higher-order filters by cascading sections and by the use of multiple-feedback paths. The all-purpose biquad was shown to be capable of realizing general biquadratic functions.

The methods discussed in this chapter will be used in Chapter 11 to obtain specific types of filters (low-pass, bandpass, Bessel, etc.) for given, practical values of gain, cutoff frequency, center frequency, quality factor, time delay, etc.

EXERCISES

10.1. For the circuit shown in Fig. Ex. 10.1 it is given that $V_o = A(s)(V_a - V_i)$, where

$$A(s) = \frac{\mu\omega_c}{s + \omega_c}, \qquad \omega_c > 0$$

p195 Fig 10.10

Show that if $\mu R_1 > R_1 + R_2$, then the transfer function V_o/V_i has a right-half plane pole, and therefore the circuit is unstable.

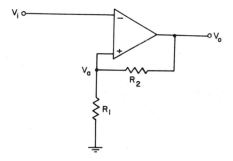

Figure Ex. 10.1.

10.2. Show that for the circuit given in Fig. Ex. 10.2, $Y = Y_1 - Y_2$. Therefore, show that if $Y = P/Q$, where Q has distinct, real, negative zeros, and the degree of P is less than or equal to 1 plus the degree of Q, then Y may be realized with RC admittances Y_1 and Y_2. (*Note:* Recall the characteristics of RC admittances from Exercise 5.20, and use a partial fraction expansion.)

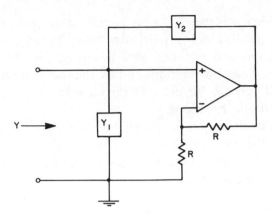

Figure Ex. 10.2.

10.3. Realize by the method of Exercise 10.2 the function

$$Y = \frac{s^4 + 1}{(s + 1)(s + 2)(s + 3)}$$

10.4. Show that the network given in Fig. Ex. 10.4 may be used with only *RC* passive elements to realize any V_2/V_1, which is a rational function with real coefficients, by obtaining, for Y_o arbitrary,

$$\frac{V_2}{V_1} = \frac{Y_2 - Y_1}{Y_3 - Y_4}$$

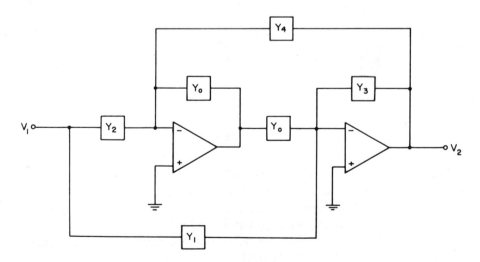

Figure Ex. 10.4.

and using a partial fraction expansion. (See Exercise 10.2.) This network is due to Lovering [L].

10.5. Using the results of Exercise 10.4, synthesize

$$\frac{V_2}{V_1} = s^4$$

10.6. The two-port network N shown in Fig. Ex. 10.6 is a *negative impedance converter* (*NIC*) if

$$Z_{in} = \frac{V_1}{I_1} = -KZ_L = \frac{KV_2}{I_2}, \qquad K > 0.$$

It is a *CNIC* if the negative sign is the result of a current inversion; that is, $V_1 = aV_2$, $I_1 = -(-bI_2)$. It is a *VNIC* if the negative sign is the result of a voltage inversion; that is, $V_1 = -aV_2$, $I_1 = -bI_2$. (In both cases, a, $b > 0$ and $K = a/b$.) Show that N is an *NIC* if and only if its transmission parameters satisfy $B = C = 0$, $AD < 0$. Show also that in this case, if $A > 0$, $D < 0$, N is a *CNIC*, and if $A < 0$, $D > 0$, N is a *VNIC*. (See Exercise 2.25.)

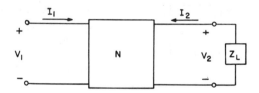

Figure Ex. 10.6.

10.7. Using the results of Exercise 10.6, show that the circuit given in Fig. Ex. 10.7 is a *CNIC* and find K.

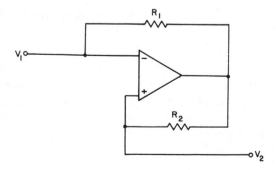

Figure Ex. 10.7.

10.8.   Show that the *RLC* driving-point function

$$Z = \frac{s^2 + s + 1}{(s + 1)(s + 2)}$$

may be realized with resistors, capacitors, and one *NIC*. (*Suggestion:* Expand Z in partial fractions.)

10.9.   A *gyrator* is a two-port network with z-parameters given by $z_{11} = z_{22} = 0$, $z_{12} = -z_{21} = -k$. The general gyrator symbol is shown in Fig. Ex. 10.9 (a), and in the 3 terminal case in (b). Show that the

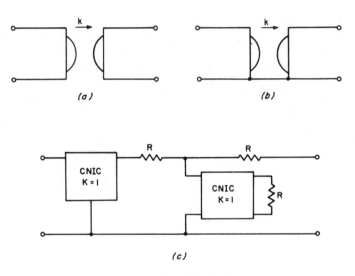

(a)   (b)

(c)

Figure Ex. 10.9.

network given in (c) is a gyrator and by Exercise 10.7 can be constructed without inductors.

10.10.   Show that a gyrator is a *positive impedance inverter*. That is, if Z_L is the load impedance, then $Z_{in} = k^2/Z_L$. Use this result to obtain $Z_{in} = Ls$ when the load is a capacitor.

10.11.   Find V_2/V_1 for the network shown in Fig. Ex. 10.11. Restrict Y_1 and Y_2 to conductances or capacitances and use the network to realize (a) a second-order normalized Bessel filter with a gain of 2, (b) a second-order normalized low-pass Butterworth filter with a gain of 1, and (c) a second-order bandpass filter with $\omega_0 = 1$ rad/s, $Q = 5$, and a gain of 4.

$$\text{Ans. } \frac{V_2}{V_1} = \frac{Y_1}{kCs(Y_1 + Y_2) + 1/k}$$

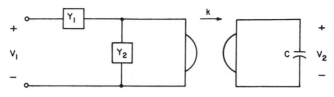

Figure Ex. 10.11.

10.12. Note that in the process of impedance scaling the network of Exercise 10.11, since k has units of ohms it must be scaled like a resistor. Using frequency and impedance scaling, obtain from part (c) of Exercise 10.11 a bandpass filter with $\omega_0 = 10{,}000$ rad/s, $Q = 5$, a gain of 4, and an output capacitor of 100 pF.

10.13. Show that the network given in Fig. Ex. 10.13 is a second-order bandpass filter, and find the transfer function V_2/V_1 for the case

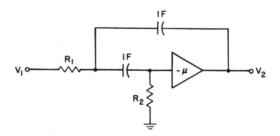

Figure Ex. 10.13.

$\mu = 3$, $R_1 = \frac{1}{8}\Omega$, and $R_2 = \frac{1}{2}\Omega$. (Note that a $VCVS$ with inverting gain can be approximated by the inverter of Fig. 10.7 if R_1 is sufficiently large that $I = V_1/R_1 \approx 0$.)

10.14. Show that in Exercise 10.13, if $\mu = 2(G_1 + 2G_2)/G_1$, then V_2/V_1 is of the form

$$H(s) = -\frac{2as}{s^2 + as + b}$$

For this case show that the network given in Fig. Ex. 10.14 is an all-pass filter, where N is the network of Exercise 10.13. For the values given in Exercise 10.13, find the all-pass transfer function.

10.15. Obtain a normalized low-pass Butterworth filter with gain $= 2$, using the infinite-gain MFB filter of Fig. 10.15 with $C_1 = 0.1F$ and $C_2 = 1F$. Denormalize the result so that $\omega_c = 10{,}000$ rad/s and the capacitances are 0.1 μF and 0.01 μF.

Ans. $R_1 = 1.3$, $R_2 = 2.6$, $R_3 = 3.85\ k\Omega$

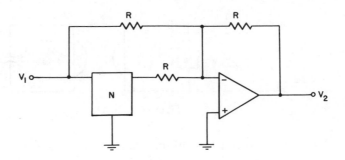

Figure Ex. 10.14.

10.16. Repeat Exercise 10.15 for a high-pass Butterworth filter using Fig. 10.16. Let $C_1 = 0.2\ \mu F$ and $C_2 = C_3 = 0.1\ \mu F$ in the denormalized case.

10.17. Repeat Exercises 10.15 and 10.16 for a 0.1 dB Chebyshev case.

10.18. Obtain an infinite-gain *MFB* second-order bandpass filter with $\omega_0 = 1$ rad/s, $Q = 10$, and $G = 6$, using Fig. 10.17, with $C_1 = C_2 = 1F$. Denormalize the result so that $f_0 = 1000$ Hz and the capacitors are 0.01 μF.

Ans. $R_1 = 26.53$, $R_2 = 0.82$, $R_3 = 318.31\ k\Omega$.

10.19. Obtain a *VCVS* second-order low-pass Chebyshev 1 dB filter with $f_c = 1000$ Hz, a gain of 2, and both capacitors 0.01 μF.

10.20. Obtain a biquad bandpass filter with $f_0 = 1000$ Hz, $Q = 50$, and $G = 10$, using Fig. 10.25 with both capacitances 0.1 μF.

Ans. $R_1 = 7.96$, $R_2 = 79.58$, all other $R_s = 1.59\ k\Omega$.

10.21. For the case $ce < 0$ discussed in Sec. 10.8, show that if in Fig. 10.26, R_8 and C_2 are interchanged and R_6 is moved to connect nodes labeled V_3 and a, then

$$\frac{V_3}{V_1} = -\frac{cs^2 + ds + e}{s^2 + as + b}$$

where if $C_1 = C_2 = G_7 = G_8 = G_9 = 1$, then $c = G_5$, $d = G_2c - G_4$, $e = G_1 - G_2G_4$, $a = G_2 - G_6$, and $b = G_3 - G_2G_6$. Thus

$$G_5 = c$$
$$G_6 = G_2 - a$$
$$G_4 = G_2c - d$$
$$G_1 = e + G_2G_4$$
$$G_3 = b + G_2G_6$$

and G_2 is arbitrary. Hence if $c > 0$, $e < 0$, we may select G_2 so that all the conductances are realizable. If $c < 0$, $e > 0$, we may realize $-V_2/V_1$ in a similar manner.

10.22. Using Fig. 10.26 or the results of Exercise 10.21, obtain active realizations of

$$\frac{V_2}{V_1} = \pm \frac{cs^2 + ds + e}{s^2 + as + b}$$

where $a = b = -d = 1$, and (a) $c = 2$, $e = 1$, (b) $c = -1$, $e = 2$, (c) $c = 0$, $e = 2$, and (d) $c = 2$, $e = 0$.

10.23. Using Fig. 10.26 and the normalized values of (10.56), obtain explicit values of the other resistances for a second-order Butterworth (a) low-pass and (b) high-pass filter, with $\omega_c = 1$ rad/s and a gain of 2. Denormalize the results so that the capacitors are each 0.01 μF and $\omega_c = 10,000$ rad/s.

10.24. Repeat Exercise 10.23 for a second-order (a) bandpass and (b) band-reject filter, with $\omega_0 = 10,000$ rad/s, $Q = 10$, and a gain of 2.

10.25. Obtain a fourth-order $\frac{1}{2}$ dB Chebyshev low-pass filter using a cascade connection of two $VCVS$ sections, having $f_c = 10,000$ Hz and $G = 4$. The capacitors are each to be 0.01 μF.

10.26. Synthesize actively the all-pass function

$$\frac{V_2}{V_1} = \frac{(s - 1)(s^2 - 2s + 2)}{(s + 1)(s^2 + 2s + 2)}$$

The capacitors to be used are each $1F$.

use

Fig 10.26

p 208~29

all purpose

biqued

11

Active Filters

11.1 GENERAL PROCEDURE

Since the specifications, either calculated or obtained from Appendices A or B, for low-pass filters are more readily available for the normalized ($\omega_c = 1$ rad/s) case, we shall find it easier to obtain the normalized network first. In this case the numbers are also more convenient since values like $1F$ or 1Ω are common. The normalized functions for the other common types of filters may be obtained by frequency transformations from those of the low-pass filters.

We should select at least one capacitance as $C = 1F$ to begin with, and any others should be standard multiples like 1, 0.1, 2, 0.47, etc. Then in the denormalization process when C is changed from $1F$ to a value like 0.01 μF, the others will change to standard values like 0.01, 0.001, 0.02, 0.0047 μF, etc.

The denormalization process is carried out, of course, by dividing the normalized capacitances by the frequency scale factor ω_c, yielding a filter with cutoff ω_c. Then dividing the capacitances and multiplying the resistances by the impedance scale factor k_i completes the process. Thus a capacitance $C = 1F$ in the normalized network becomes $C = 1/\omega_c k_i$ in the denormalized network, and any other capacitance C'_k in

normalized $C = 1F$ $\longrightarrow$ denormalized $C = \dfrac{1}{w_c k_i}$

the normalized network becomes

$$C_k = \frac{C'_k}{\omega_c k_i} = C'_k C \qquad (11.1)$$

in the denormalized network. Since C is usually selected to be some standard value (and thus C'_k should be a standard multiple), then k_i is found from

$$k_i = \frac{1}{\omega_c C} = \frac{1}{2\pi f_c C} \qquad \boxed{\begin{array}{c} \omega_c = 2\pi f_c \\ f_c = \dfrac{\omega_c}{2\pi} \end{array}} \quad (11.2)$$

A good rule of thumb is to select a standard value of C near $10/f_c$ μF.

If the denormalized resistances obtained are not suitable, it is well to remember that they may all be multiplied by a common factor provided the capacitances are divided by the same factor. In general, feedback resistances may be much larger relative to the other resistances and still provide good circuit performance.

In the following sections we shall obtain explicit element values in a number of cases of various types of active filters. For many complete designs of low-pass, high-pass, bandpass, band-reject, all-pass, and constant time delay filters available for practical element values for frequencies of interest from 1 to 10^6 Hz, the reader is referred to reference [HJ]. The design data is given in a form which requires no computations whatsoever. A more complete source, requiring minimal computations, is reference [JH].

11.2 INFINITE-GAIN MULTIPLE-FEEDBACK LOW-PASS FILTERS

As was noted earlier in Sec. 10.5, the infinite-gain *MFB* filter of Fig. 10.15, repeated here as Fig. 11.1, is a low-pass filter, having transfer

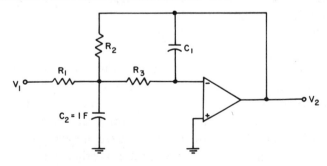

Figure 11.1. An infinite-gain *MFB* low-pass filter.

function

$$\frac{V_2}{V_1} = \frac{-Gb_0}{s^2 + b_1 s + b_0}$$

where by (10.22),

$$\begin{cases} b_0 = \dfrac{G_2 G_3}{C_1} \\[2mm] b_1 = G_1 + G_2 + G_3 \\[2mm] G = \dfrac{G_1}{G_2} \end{cases} \qquad (11.3)$$

(We have taken $C_2 = 1F$ for convenience in the normalized case.)

To design a second-order low-pass filter, or section, we select b_0 and b_1 from Appendix A, or derive them from the given specifications, for a normalized cutoff $\omega_c = 1$ rad/s. For the given G we determine the normalized conductances from (11.3), which become, when the first and third are substituted into the second,

$$b_1 = GG_2 + G_2 + \frac{C_1 b_0}{G_2}$$

or

$$(1 + G)G_2^2 - b_1 G_2 + C_1 b_0 = 0$$

Solving for G_2 we have

$$G_2 = \frac{b_1 + \sqrt{b_1^2 - 4C_1 b_0(1 + G)}}{2(1 + G)} \qquad (11.4)$$

We have chosen the plus sign on the radical but we could have chosen also the negative sign.

For a given set of values b_0, b_1, and G we must choose C_1 in (11.4) so that G_2 is real. The other conductances are given from the first and third of (11.3) by

$$G_1 = GG_2, \qquad G_3 = \frac{C_1 b_0}{G_2} \qquad (11.5)$$

As an example, if we wish a low-pass Butterworth filter with $G = 10$ and $f_c = 10{,}000$ Hz, we have $b_0 = 1$, $b_1 = \sqrt{2}$, and from (11.4),

$$G_2 = \frac{\sqrt{2} + \sqrt{2 - 44C_1}}{22}$$

Let us choose $C_1 = 0.01F$ so that $G_2 = 0.1211$. From (11.5) we have $G_1 = 1.211$ and $G_3 = 0.0826$. If we wish C_2 to be 0.1 μF, then by (11.1)

$$C_1 = 0.01 \times 10^{-7}\, F = 0.001\ \mu F$$

and

$$k_i = \frac{1}{(2\pi)(10^4)(10^{-7})} = 159.15$$

Therefore the denormalized resistances are

$$R_1 = \frac{k_i}{G_1} = 131.42\Omega$$

$$R_2 = \frac{k_i}{G_2} = 1.314\ \text{k}\Omega$$

$$R_3 = \frac{k_i}{G_3} = 1.927\ \text{k}\Omega$$

The infinite-gain *MFB* filter achieves an inverting-gain low-pass characteristic with a minimal number of circuit elements. It also has the advantages of low output impedance and good stability [Hu-1].

11.3 *VCVS* LOW-PASS FILTERS

The second-order *VCVS* low-pass filter of Fig. 10.19 achieves the transfer function

$$\frac{V_2}{V_1} = \frac{Gb_0}{s^2 + b_1 s + b_0} \qquad (11.6)$$

provided (10.31) holds. Taking $C_2 = 1F$, Fig. 10.19 is as shown in Fig. 11.2 and (10.31) becomes

$$G = \mu = 1 + \frac{R_4}{R_3} \geq 1$$

$$b_0 = \frac{G_1 G_2}{C_1} \qquad (11.7)$$

$$b_1 = G_1 + G_2 + (1 - \mu)\frac{G_2}{C_1}$$

Substituting the first two of (11.7) into the third yields

$$G_1^2 - b_1 G_1 + b_0(C_1 + 1 - G) = 0$$

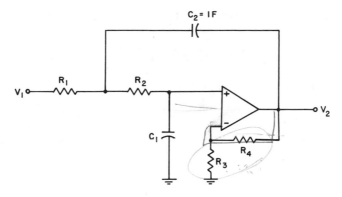

Figure 11.2. A second-order *VCVS* low-pass filter.

from which we have, with (11.7),

$$G_1 = \frac{b_1 + \sqrt{b_1^2 - 4b_0(C_1 + 1 - G)}}{2}$$

$$G_2 = \frac{C_1 b_0}{G_1} \tag{11.8}$$

(Since for $G > C_1 + 1$ the discriminant exceeds b_1^2, we have chosen the positive sign on the radical.) Also, for minimum *dc* offset we have by (10.32),

p203

$$\left(\begin{array}{l} R_4 = G(R_1 + R_2) \\[2mm] R_3 = \dfrac{G(R_1 + R_2)}{G - 1}, \quad G \neq 1 \end{array}\right. \tag{11.9}$$

If $G = 1$, then we may take $R_4 = 0$ and $R_3 = \infty$. (We could for minimum *dc* offset in this case, take $R_4 = R_1 + R_2$.)

Therefore, given the type of filter (b_0 and b_1) and the gain G, we select C_1 so that G_1 is real, requiring from (11.8) that

$$0 < C_1 \leq G - 1 + \frac{b_1^2}{4b_0} \tag{11.10}$$

We then find the resistances from (11.8) and (11.9) and denormalize to the required C_2 and ω_c.

As an example, suppose we wish a Butterworth filter ($b_0 = 1$, $b_1 = \sqrt{2}$) with $G = 1$, $\omega_c = 10^5$ rad/s, and $C_2 = 0.01$ μF. By (11.10) we have $C_1 \leq \frac{1}{2}$. Choosing $C_1 = \frac{1}{2}F$, we have from (11.8), $R_1 = R_2 =$

$\sqrt{2}\,\Omega$. The impedance scale factor is given by

$$k_i = \frac{1}{(10^5)(10^{-8})} = 1000$$

so that the denormalized resistances are

$$R_1 = R_2 = \sqrt{2}\,(1000)\Omega = 1.414\ k\Omega$$

The denormalized C_1 is $0.5(0.01) = 0.005\ \mu F$. The circuit is shown in Fig. 11.3.

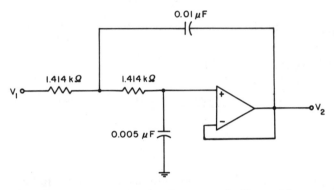

Figure 11.3. A second-order Butterworth filter with $\omega_c = 10^5$ rad/s.

The *VCVS* low-pass filter is one of the more popular configurations with noninverting gain. It requires a minimal number of elements (only one more than the infinite-gain *MFB* filter of the previous section), has a relative ease of adjustment of characteristics, and a low output impedance [Hu-1]. A definite advantage is the ability to set the gain precisely by setting the gain of the *VCVS* with a potentiometer in lieu of resistors R_3 and R_4.

11.4 BIQUAD LOW-PASS FILTERS

The biquad of Fig. 10.25, redrawn as shown in Fig. 11.4 for

$$C = G_4 = 1 \qquad (11.11)$$

yields the low-pass function of (11.6) if

$$G_1 = Gb_0, \qquad G_2 = b_1, \qquad G_3 = b_0 \qquad (11.12)$$

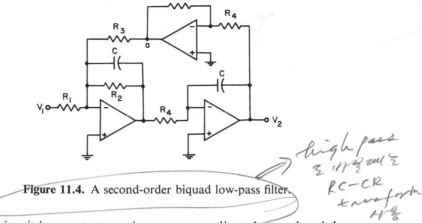

Figure 11.4. A second-order biquad low-pass filter.

[handwritten: high pass ... $\frac{3}{2}$ ... $RC-CR$ transform ... $H\frac{s}{\omega}$ ... $R \to \frac{1}{C}$ ... $C \to \frac{1}{R}$ ... $(P \, 83)$ ... $\frac{?}{\omega^2}$]

Thus the circuit is easy to tune since we may adjust the passband characteristics (b_0 and b_1) by varying G_2 and G_3, and we may set the gain by varying G_1. If we want an inverting gain we may take the output at node a.

The biquad requires more elements than either the infinite-gain MFB or the VCVS low-pass filters of the previous two sections. However, this disadvantage is offset by the ease of tuning the biquad and its relatively good stability [T-3]. These properties also make it comparatively easy to cascade several biquad sections to obtain higher-order filters.

11.5   STATE VARIABLE LOW-PASS FILTERS

In this section we shall consider a method equivalent to the state variable method [KHN], referred to in Sec. 10.7, for obtaining higher order low-pass filters. Let us consider the network of Fig. 11.5, where each resonator N_i has transfer function $H_i = V_{i+1}/V_i$, $i = 1, 2, \ldots, n-1$.

Evidently the output of N_1 is $V_2 = H_1V_1$, of N_2 is $V_3 = H_1H_2V_1$, etc., and of N_{n-1} is

$$V_{out} = V_n = H_1H_2 \cdots H_{n-1}V_1 \qquad (11.13)$$

At node a we have

$$-G_0V_{in} - (s + G_1)V_1 - G_2H_1V_1 - G_3H_1H_2V_1 - \cdots$$
$$- G_nH_1H_2 \cdots H_{n-1}V_1 = 0$$

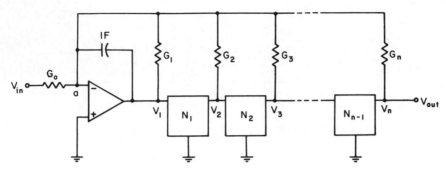

Figure 11.5. A general higher-order circuit.

or by (11.13),

$$\frac{V_{out}}{V_{in}} = \frac{-G_0 H_1 H_2 \cdots H_{n-1}}{s + G_1 + G_2 H_1 + G_3 H_1 H_2 + \cdots + G_n H_1 H_2 \cdots H_{n-1}} \quad (11.14)$$

This would be the transfer function of an nth-order low-pass filter if $H_i = 1/s$ for $i = 1, 2, \ldots, n - 1$, and the network would be easy to tune since G_i is the coefficient of s^{n-i} when (11.14) is simplified. The difficulty however, is that a simple integrator such as Fig. 10.13 yields

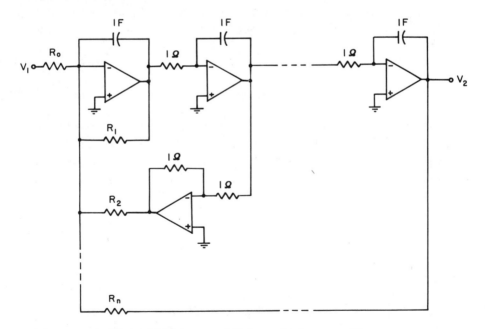

Figure 11.6. A general higher-order low-pass filter.

$-1/s$ as a transfer function. This difficulty could be overcome by placing an inverter in series with every other feedback conductance starting with G_2. The resulting circuit is shown in Fig. 11.6, and the transfer function is given by

$$\frac{V_2}{V_1} = \frac{(-1)^n G_0}{s^n + G_1 s^{n-1} + \cdots + G_{n-1}s + G_n}; \qquad n = 1, 2, 3, \ldots \qquad (11.15)$$

As an example, if $n = 2$, the result is identical to Fig. 10.25, the low-pass biquad. For $n = 3$, the transfer function is

$$\frac{V_2}{V_1} = \frac{-G_0}{s^3 + G_1 s^2 + G_2 s + G_3} \qquad (11.16)$$

and the circuit is shown in Fig. 11.7.

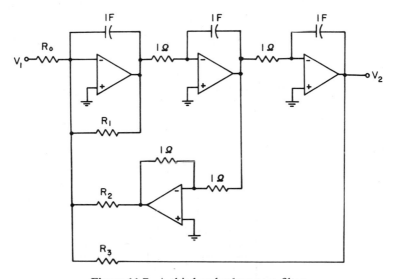

Figure 11.7. A third-order low-pass filter.

11.6 LOW-PASS ELLIPTIC FILTERS USING BIQUADS

In the case of a second-order low-pass elliptic filter, as was shown in Sec. 6.4, the transfer function is given, for inverting gain, by

$$\frac{V_2}{V_1} = \frac{-K(s^2 + a_1)}{s^2 + b_1 s + b_0} \qquad (11.17)$$

where the gain is

$$G = \frac{Ka_1}{b_0} \tag{11.18}$$

This function may be realized by the biquad of Fig. 10.26, if we choose in (10.47), $d = 0$, $c = K$, $e = a_1K$, $a = b_1$, and $b = b_0$. Selecting the normalized values $C_1 = C_2 = G_7 = G_8 = G_9 = 1$, and $G_6 = 0$, we have from (10.57),

$$G_1 = b_1K = \frac{b_1b_0G}{a_1}$$

$$G_2 = b_1$$

$$G_3 = b_0 \tag{11.19}$$

$$G_4 = K = \frac{b_0G}{a_1}$$

$$G_5 = \frac{a_1K}{b_0} = G$$

The circuit is shown in Fig. 11.8.

Another elliptic filter realization, due to Kerwin, Huelsman, and Newcomb [KHN], is given in Exercise 11.9.

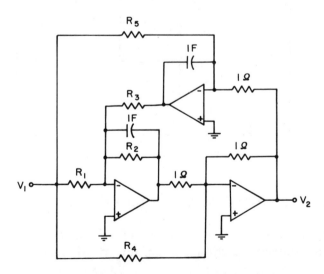

Figure 11.8. A second-order low-pass elliptic filter.

11.7 INFINITE-GAIN *MFB* HIGH-PASS FILTERS

An infinite-gain *MFB* high-pass filter of second order was obtained in Sec. 10.5 and realizes the general function

$$\frac{V_2}{V_1} = \frac{-Gs^2}{s^2 + a_1 s + a_0} \tag{11.20}$$

The circuit is shown in Fig. 11.9 and if $C_1 = 1F$, by (10.24) we have

$$a_0 = \frac{G_1 G_2}{C_2}$$

$$a_1 = \frac{G_2(C_2 + 2)}{C_2} \tag{11.21}$$

$$G = \frac{1}{C_2}$$

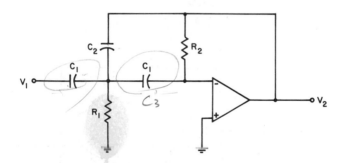

Figure 11.9. An infinite-gain *MFB* high-pass filter.

Solving (11.21) for the normalized circuit elements we have

$$C_2 = \frac{1}{G}$$

$$G_1 = \frac{a_0(2G + 1)}{a_1 G} \tag{11.22}$$

$$G_2 = \frac{a_1}{(2G + 1)}$$

Thus if $1/G$ is a standard capacitance multiple like $1, 2, \frac{1}{2}$, etc., then C_2 is a standard value and G_1 and G_2 are calculated from (11.22). The denormalization process is identical to that in the low-pass case.

The relative merits of the infinite-gain *MFB* high-pass filter are the same as those of its low-pass counterpart listed earlier in Sec. 11.2.

11.8 *VCVS* HIGH-PASS FILTERS

Choosing $C_1 = C_2 = C$ in Fig. 10.21, we have the *VCVS* high-pass filter of second order shown in Fig. 11.10. It realizes the general function

$$\frac{V_2}{V_1} = \frac{Gs^2}{s^2 + a_1 s + a_0} \qquad (11.23)$$

where by (10.36), if $C = 1F$, we have

$$G = 1 + \frac{R_4}{R_3} \geq 1$$
$$a_0 = G_1 G_2 \qquad (11.24)$$
$$a_1 = 2G_2 + G_1(1 - G)$$

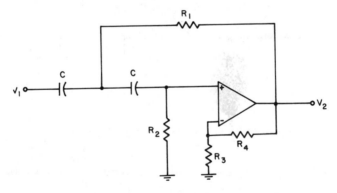

Figure 11.10. A second-order *VCVS* high-pass filter.

For minimum *dc* offset, by (10.32), we have

$$R_4 = GR_2$$
$$R_3 = \frac{GR_2}{(G - 1)}, \qquad G \neq 1 \qquad (11.25)$$

If $G = 1$, we may take $R_4 = 0$ and $R_3 = \infty$. (For minimum *dc* offset in this case we would take $R_4 = R_2$.)

From the second and third of (11.24) we have

$$2G_2^2 - a_1G_2 + a_0(1 - G) = 0$$

or

$$G_2 = \frac{a_1 + \sqrt{a_1^2 + 8a_0(G - 1)}}{4}$$

$$G_1 = \frac{a_0}{G_2}$$

(11.26)

Thus for a given type (a_0 and a_1) and gain G, we may find the normalized conductances. The denormalization is carried out as in the low-pass case.

The advantages of the *VCVS* high-pass filter are the same as those of the *VCVS* low-pass filter given in Sec. 11.3.

As an example, suppose we want a second-order high-pass 1 dB Chebyshev filter with $G = 2$ and $\omega_c = 10,000$ rad/s, using 0.01 μF capacitors. By Appendix A and (4.9) we have

$$a_0 = \frac{1}{1.10251} = 0.90702$$

$$a_1 = \frac{1.09773}{1.10251} = 0.99566$$

Thus the normalized conductances are

$$G_2 = \frac{0.99566 + \sqrt{(0.99566)^2 + 8(0.90702)(1)}}{4} = 0.96688$$

and

$$G_1 = \frac{0.90702}{0.96688} = 0.93809$$

The impedance scale factor is given by

$$k_i = \frac{1}{\omega_c C} = \frac{1}{(10^4)(10^{-8})} = 10^4$$

Thus the denormalized resistances are $R_1 = 10.660\ k\Omega$, $R_2 = 10.343\ k\Omega$, and $R_3 = R_4 = 2R_2 = 20.686\ k\Omega$.

11.9 BIQUAD HIGH-PASS FILTERS

A second-order biquad high-pass filter may be realized with Fig. 10.26 if $d = e = 0$. The transfer function with inverting gain G is given by (11.20), so that in (10.46) we have $c = G$, $a = a_1$, and $b = a_0$. Taking $G_6 = 0$ and the normalized values of (10.56), we have by (10.57),

$$G_1 = a_1 G$$
$$G_2 = a_1$$
$$G_3 = a_0 \tag{11.27}$$
$$G_4 = G$$
$$G_5 = 0$$

The circuit is shown in Fig. 11.11.

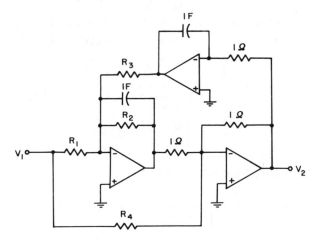

Figure 11.11. A biquad high-pass filter.

An elliptic high-pass filter may also be obtained from Fig. 10.26. A second-order elliptic high-pass function may be obtained from the low-pass case of (11.17) by replacing s by $1/s$. The result after simplification is

$$\frac{V_2}{V_1} = \frac{-G\left(s^2 + \dfrac{1}{a_1}\right)}{s^2 + \dfrac{b_1}{b_0}s + \dfrac{1}{b_0}} \tag{11.28}$$

so that the gain, as in the low-pass case, is inverting and given by G.

Since (11.28) is of the exact form as (11.17), it may be realized also by Fig. 11.8, with the normalized values given by

$$G_1 = \frac{Gb_1}{b_0}$$

$$G_2 = \frac{b_1}{b_0}$$

$$G_3 = \frac{1}{b_0} \qquad\qquad (11.29)$$

$$G_4 = G$$

$$G_5 = \frac{Gb_0}{a_1}$$

Another high-pass biquad circuit, as was pointed out in Sec. 10.7, is given by Fig. 10.23 with the output V_2 taken at node 4. The transfer function is given by (11.20) if

$$G_1 = G, \qquad G_2 = a_1, \qquad G_3 = a_0 \qquad (11.30)$$

11.10 INFINITE-GAIN *MFB* BANDPASS FILTERS

The second-order infinite gain *MFB* bandpass filter was given earlier in Fig. 10.17, which is repeated here as Fig. 11.12. It realizes the function (10.25), which normalized to $\omega_0 = 1$ rad/s ($B = 1/Q$) is given by

$$\frac{V_2}{V_1} = \frac{-\dfrac{G}{Q}s}{s^2 + \dfrac{1}{Q}s + 1} \qquad (11.31)$$

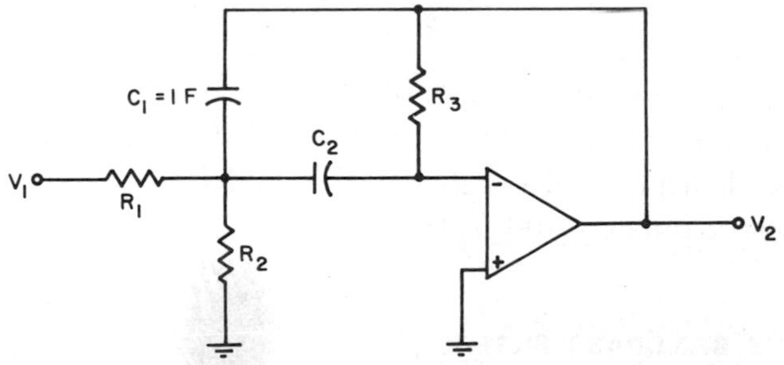

Figure 11.12. A second-order infinite-gain *MFB* bandpass filter.

By (10.26), for this case, we have

$$\frac{1}{Q} = (1 + C_2)\frac{G_3}{C_2}$$

$$1 = \frac{G_3(G_1 + G_2)}{C_2} \qquad (11.32)$$

$$G = \frac{G_1 C_2}{G_3(1 + C_2)}$$

where we have taken $C_1 = 1F$.

Solving (11.32), for C_2 arbitrary, yields the normalized conductances

$$G_1 = \frac{G}{Q}$$

$$G_2 = Q(1 + C_2) - \frac{G}{Q} \qquad (11.33)$$

$$G_3 = \frac{C_2}{Q(1 + C_2)}$$

Therefore, given G and Q ($\omega_0 = 1$ rad/s), we select C_2 from the second of (11.33) so that $G_2 > 0$, and find the normalized conductances. Then we denormalize the network for a given value of C_1 to the desired center frequency ω_0. As before, as in the low- and high-pass cases, the impedance scale factor is given by

$$k_i = \frac{1}{\omega_0 C_1} \qquad (11.34)$$

The denormalized value of C_2 is

$$C_2 = \frac{C_2'}{\omega_0 k_i} = C_2' C_1 \qquad (11.35)$$

where C_2' is the normalized value of C_2.

The infinite-gain MFB filter yields inverting gain, has a minimal number of elements, and is capable, for moderate gains, of values of Q up to approximately 10 [GTH].

11.11 *VCVS* BANDPASS FILTERS

To obtain a *VCVS* bandpass filter, we add a capacitor from node a to ground and select the other elements of Fig. 10.18 as shown in

Fig. 11.13 [KH]. The circuit realizes the normalized function

$$\frac{V_2}{V_1} = \frac{\dfrac{G}{Q}s}{s^2 + \dfrac{1}{Q}s + 1} \tag{11.36}$$

for $C = 1F$, with

$$\frac{G}{Q} = \mu G_1$$

$$\frac{1}{Q} = G_1 + G_2(1 - \mu) + 2G_3 \tag{11.37}$$

$$1 = G_3(G_1 + G_2)$$

where

$$\mu = 1 + \frac{R_5}{R_4}$$

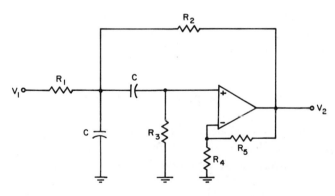

Figure 11.13. A second-order *VCVS* bandpass filter.

Taking μ as arbitrary, we may solve for the normalized conductances, obtaining after some simplification, for $\mu > 1$,

$$G_1 = \frac{G}{\mu Q}$$

$$G_2 = \frac{G - 1 + \sqrt{(G-1)^2 + 8(\mu-1)Q^2}}{2(\mu-1)Q} - \frac{G}{\mu Q} \tag{11.38}$$

$$G_3 = \frac{1}{(G_1 + G_2)}$$

(If $\mu = 1$, we may show from (11.37) that $0 < G < 1$, which is a case we shall omit.) Finally, let us take $\mu = 2$, requiring in Fig. 11.13 that $R_4 = R_5$, and for minimum dc offset,

$$R_4 = R_5 = 2R_3 \qquad (11.39)$$

In this case (11.38) becomes

$$G_1 = \frac{G}{2Q}$$

$$G_2 = \frac{(\sqrt{(G-1)^2 + 8Q^2} - 1)}{2Q} \qquad (11.40)$$

$$G_3 = \frac{1}{(G_1 + G_2)}$$

Thus we may choose μ in (11.38) so that $G_2 > 0$ and find the normalized conductances. However, since we shall always have $Q^2 > \frac{1}{8}$, then $G_2 > 0$ in (11.40) and hence (11.39) and (11.40) may be used.

The *VCVS* bandpass filter has the advantages of the *VCVS* low-pass network, enumerated previously. For moderate gains it can achieve values of Q up to 25 or so [JH].

11.12 A POSITIVE FEEDBACK BANDPASS FILTER

A second-order bandpass filter which is capable of Qs up to about 50 is the circuit [GTH] of Fig. 11.14. We shall refer to it as a *positive feedback* circuit because the signal fed back through R_3 is an uninverted signal.

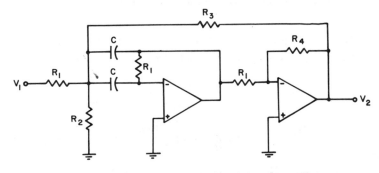

Figure 11.14. A positive feedback bandpass filter.

Taking $C = 1F$, analysis of Fig. 11.14 shows that it realizes the normalized second-order bandpass function with

$$\frac{1}{Q} = 2G_1 - \frac{G_1 G_3}{G_4}$$
$$1 = G_1(G_1 + G_2 + G_3) \qquad (11.41)$$
$$\frac{G}{Q} = \frac{G_1^2}{G_4}$$

Taking G_4 as arbitrary, we may solve (11.41) for the conductances, obtaining

$$G_1 = \sqrt{GG_4/Q}$$
$$G_2 = \sqrt{Q/GG_4} - \sqrt{GG_4/Q} + \sqrt{G_4/QG} - 2G_4 \qquad (11.42)$$
$$G_3 = 2G_4 - \sqrt{G_4/QG}$$

For a given Q and G, we may take G_4 sufficiently small so that the conductances are positive. To see this, we note that $G_3 \geq 0$ if by the third of (11.42) we have

$$G_4 \geq \frac{1}{4QG} \qquad (11.43)$$

Taking $G_4 = 1/4QG$ we have $G_3 = 0$ (the cascade case), and from the second of (11.42),

$$G_2 = 2Q - \frac{1}{2Q}$$

Thus $G_2 \geq 0$ if $Q \geq \frac{1}{2}$. Finally, dG_2/dG_4 at $G_4 = 1/4QG$ is given by

$$\frac{dG_2}{dG_4} = -(4Q^2 G + G + 1) < 0$$

Therefore for $Q > \frac{1}{2}$, a value of $G_4 > 1/4QG$ exists so that both G_2 and G_3 are positive.

11.13 A BIQUAD BANDPASS FILTER

For the second-order bandpass filter we shall use the biquad of Fig. 11.15, which is that of Fig. 10.25 with the integrator and inverter inter-

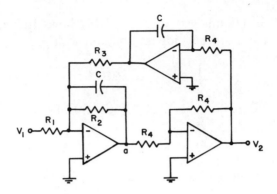

Figure 11.15. A biquad bandpass filter.

changed. Taking the normalized values $C = R_4 = 1$, we have

$$\frac{V_2}{V_1} = \frac{G_1 s}{s^2 + G_2 s + G_3} \qquad (11.44)$$

Therefore the circuit is a normalized bandpass filter with a given Q and G provided

$$G_1 = \frac{G}{Q}$$

$$G_2 = \frac{1}{Q} \qquad (11.45)$$

$$G_3 = 1$$

To obtain the denormalized case we may match the transfer function of Fig. 11.15, without the normalized case $C_1 = R_4 = 1$, with the denormalized transfer function,

$$\frac{V_2}{V_1} = \frac{\dfrac{G\omega_0 s}{Q}}{s^2 + \dfrac{\omega_0}{Q} s + \omega_0^2}$$

This results in

$$R_1 = \frac{Q}{G\omega_0 C}$$

$$R_2 = \frac{Q}{\omega_0 C} \qquad (11.46)$$

$$R_3 = \frac{1}{\omega_0^2 R_4 C^2}$$

where R_4 and C are arbitrary. Thus we may obtain the normalized network using (11.45) and denormalize for a given C and ω_0, or we may use (11.46) directly to obtain the denormalized network. In the latter case, if R_4 is selected equal to R_3, then by the last of (11.46) we have

$$R_3 = R_4 = \frac{1}{\omega_0 C} \tag{11.47}$$

If we wish an inverting gain, we may take the output in Fig. 11.15 at node a. The equations for the conductances are unchanged.

The biquad circuit is easily tuned, as is seen from (11.46). The center frequency may be adjusted by varying R_3, Q may be adjusted by varying R_2, and the gain may be adjusted by varying R_1.

The biquad circuit is capable of attaining high values of Q, say up to 100, and is a much more stable network than those of the previous three sections. It is therefore more amenable to the cascade method of synthesis. For this purpose we have to match the general transfer function (11.44) to each second-order section.

11.14 A MULTIPLE-RESONATOR BANDPASS FILTER

Let us consider in this section a multiple-feedback circuit, shown in Fig. 11.16, which is very similar to the state-variable structure of Fig. 11.5. If the resonators N_i are all identical bandpass sections, as suggested by Hurtig [Hg], a $2n$th-order bandpass filter results [JHI].

Analysis of Fig. 11.16 yields

$$\frac{V_2}{V_1} = \frac{-a_0 H_1 H_2 \cdots H_n}{1 + a_1 H_1 + a_2 H_1 H_2 + \cdots + a_n H_1 H_2 \cdots H_n} \tag{11.48}$$

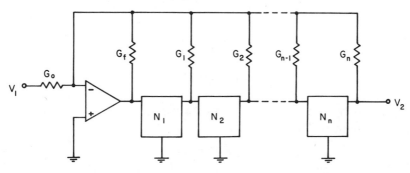

Figure 11.16. A general multiple-feedback network.

where $H_i = V_{out}/V_{in}$ is the transfer function of resonator N_i, $i = 1, 2,$ $\ldots, n$, and

$$a_i = G_i/G_f, \qquad i = 0, 1, \ldots, n \qquad (11.49)$$

It is understood that the resonators are RC, op amp sections with proper isolation provided by the op amps.

Suppose the resonators are identical bandpass sections with

$$H_i = \frac{V_{out}}{V_{in}} = \frac{K_1 s/Q_1}{s^2 + \dfrac{1}{Q_1} s + 1}, \qquad i = 1, 2, \ldots, n \qquad (11.50)$$

(The center frequency is normalized to 1 rad/s, the gain is K_1, and the quality factor is Q_1.) Then (11.48) becomes

$$\frac{V_2}{V_1} = \frac{-a_0 K_1^n s^n / Q_1^n}{D_1(s)} \qquad (11.51)$$

where

$$D_1(s) = \left(s^2 + \frac{1}{Q_1} s + 1\right)^n + \sum_{i=1}^{n} \frac{a_i K_1^i s^i \left(s^2 + \dfrac{1}{Q_1} s + 1\right)^{n-i}}{Q_1^i} \qquad (11.52)$$

Applying the binomial expansion to $D_1(s)$ we have

$$D_1(s) = \sum_{i=0}^{n} \binom{n}{i}(s^2 + 1)^{n-i}(s/Q)^i$$

$$+ \sum_{i=1}^{n} \sum_{j=0}^{n-i} a_i K_1^i \binom{n-i}{j}(s^2 + 1)^{n-i-j}(s/Q_1)^{i+j}$$

which may be rewritten in the form

$$D_1(s) = (s^2 + 1)^n + \sum_{i=1}^{n}\left[\binom{n}{i} + \sum_{j=1}^{i} a_j K_1^j \binom{n-j}{i-j}\right](s^2 + 1)^{n-i}(s/Q_1)^i \qquad (11.53)$$

where by definition

$$\binom{n}{r} = \frac{n!}{r!(n-r)!}$$

The general $2n$th-order normalized bandpass function is given, for inverting gain G, by

$$\frac{V_2}{V_1} = \frac{-Gb_0}{P_n(S)} \qquad (11.54)$$

where

$$P_n(S) = S^n + b_{n-1}S^{n-1} + \cdots + b_1S + b_0 \qquad (11.55)$$

and

$$S = \frac{Q(s^2 + 1)}{s} \qquad (11.56)$$

Substituting (11.55) and (11.56) into (11.54) results in

$$\frac{V_2}{V_1} = \frac{-Gb_0 s^n / Q^n}{D_2(s)} \qquad (11.57)$$

where

$$D_2(s) = \sum_{i=0}^{n} \frac{b_{n-i}s^i(s^2 + 1)^{n-i}}{Q^i}, \qquad b_n = 1 \qquad (11.58)$$

The multiple-feedback filter is obtained by matching (11.57) and (11.58) with (11.51) and (11.53), resulting in

$$\frac{a_0 K_1^n}{Q_1^n} = \frac{Gb_0}{Q^n} \qquad (11.59)$$

and

$$\binom{n}{i}\frac{1}{Q_1^i} + \sum_{j=1}^{i} a_j K_1^j \binom{n-j}{i-j}\frac{1}{Q_1^i} = \frac{b_{n-i}}{Q^i}, \qquad i = 1, 2, \ldots, n \quad (11.60)$$

The approach used in reference [JHI] is to make the assignments

$$K_1 = 1/c, \qquad Q_1 = Q/c \qquad (11.61)$$

where c is arbitrary, to be used, hopefully, to make the a_i nonnegative. Substituting (11.61) into (11.59) and (11.60) yields

$$a_0 = Gb_0 \qquad (11.62)$$
$$a_1 = b_{n-1} - cn \qquad (11.63)$$

and

$$a_i = b_{n-i} - \binom{n}{i}c^i - \sum_{j=1}^{i-1}\binom{n-j}{i-j}a_j c^{i-j}, \qquad i = 2, 3, \ldots, n \quad (11.64)$$

Thus for a chosen value of c, we may find a_0, a_1, and by (11.64), successively $a_2, a_3, \ldots, a_n$.

We may also solve (11.64) explicitly for a_i for which it may be shown (see Exercise 11.24) that

$$a_i = (-1)^i \sum_{j=0}^{i} (-1)^j \binom{n-j}{i-j} c^{i-j} b_{n-j}, \qquad i = 1, 2, \ldots, n \qquad (11.65)$$

The range on c for nonnegative a_i depends on the low-pass prototype coefficients b_i. It is interesting to note from (11.65) that

$$a_n = P_n(-c)$$

where $P_n(s)$ is the denominator polynomial of (11.55). Since $P_n(s)$ is strictly Hurwitz, all its zeros occur in the left-half of the s-plane. Also $P_n(0) > 0$. Therefore, if $-\sigma$ ($\sigma > 0$) is the real zero of $P_n(s)$ nearest the origin, then any value of c on $0 < c < \sigma$ results in $a_n > 0$. By (11.63), $a_1 \geq 0$ if

$$0 < c \leq \frac{b_{n-1}}{n} \qquad (11.66)$$

Other bounds on c arise from (11.65). The reader may verify that for

$$c = \frac{b_{n-1}}{2n} \qquad (11.67)$$

the center of the interval (11.66), all the a_i are positive for $n = 1, 2, 3,$ and 4 in the Butterworth and the $0.1, 0.5, 1, 2,$ and 3 dB Chebyshev cases. This result probably holds for a general n as well. Also we might note that the a_i are nonnegative if $c = b_{n-1}/n$, which results in $a_1 = 0$ and eliminates resistor R_1.

The advantages of the *multiple resonator* arrangement of Fig. 11.16 are that the gain G can be set with resistor R_0, given from (11.49) and (11.62) by

$$R_0 = \frac{R_f}{Gb_0} \qquad (11.68)$$

the b_i of the low-pass prototype are determined by the feedback network of resistors, as is seen by (11.63) and (11.64), and the center frequency and Q are determined by the resonators. Thus we have a multiple-feedback system which retains the advantage of separately tuneable resonators, as in the cascade case.

The resonators in Fig. 11.14 must have noninverting gain K_1 and quality factor Q_1, given by (11.61). Since c is normally a number less than 1, we see that the resonators must be capable of providing $Q_1 > Q$.

As an example, suppose we want an eighth-order ($n = 4$) 1 dB Chebyshev bandpass filter with $f_0 = 1000$ Hz, $Q = 5$, and $G = 1$. Selecting c from (11.67) we have from (11.61), (11.62), and (11.65) for the general eighth-order case,

$$K_1 = \frac{8}{b_3}, \qquad Q_1 = \frac{8Q}{b_3}$$

$$a_0 = Gb_0$$

$$a_1 = \frac{b_3}{2}$$

$$a_2 = b_2 - \frac{9b_3^2}{32}$$

$$a_3 = b_1 - \frac{b_2 b_3}{4} + \frac{5b_3^3}{128}$$

$$a_4 = b_0 - \frac{b_1 b_3}{8} + \frac{b_2 b_3^2}{64} - \frac{7b_3^4}{4096}$$

From Appendix A we may find the b_i for the fourth-order 1 dB Chebyshev case, from which we obtain $K_1 = 8.39621$, $Q_1 = 41.98103$, $a_0 = 0.27563$, $a_1 = 0.47641$, $a_2 = 1.19859$, $a_3 = 0.43008$, and $a_4 = 0.20640$. Since $Q_1 < 50$ we may use the positive feedback resonator of Sec. 11.12, for which, choosing $G_4 = 0.5$, we have by (11.42), $G_1 = 0.3162$, $G_2 = 1.8841$, and $G_3 = 0.9623$. Selecting capacitances of 0.01 μF, the impedance scale factor is $k_i = 1/(2\pi)(10^3)(10^{-8}) = 1.5915 \times 10^4$. This results in resonator resistances of $R_1 = 50.33$, $R_2 = 8.45$, $R_3 = 16.54$, and $R_4 = 31.83 \, k\Omega$. Selecting $R_f = 1\Omega$ we have from $G_i = a_i$ the feedback resistances, $R_i = k_i/a_i$, given by $R_0 = 57.74$, $R_1 = 33.41$, $R_2 = 13.28$, $R_3 = 37.01$, and $R_4 = 77.11 \, k\Omega$. The denormalized value of R_f is k_i, or 15.92 $k\Omega$. The actual response obtained is shown in Fig. 11.17.

Alternately, to obtain the denormalized feedback resistances, we may choose a denormalized value of R_f and use $R_i = R_f/a_i$ for the other values. For example, if we choose $R_f = k_i = 15.915 \, k\Omega$ at first, then the other feedback resistances are obtained directly, and of course, are the same as those calculated earlier.

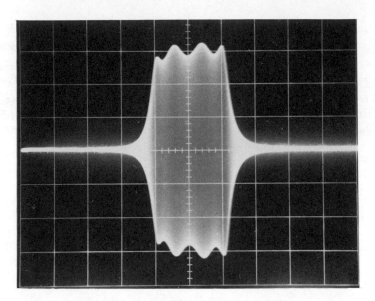

Figure 11.17. An eighth-order Chebyshev bandpass response.

11.15   A *VCVS* BAND-REJECT FILTER

A *VCVS* filter which realizes the second-order band-reject function

$$\frac{V_2}{V_1} = \frac{G(s^2 + \omega_0^2)}{s^2 + Bs + \omega_0^2} \qquad (11.69)$$

is shown in Fig. 11.18 [I]. Let us consider the normalized case, $\omega_0 = 1$

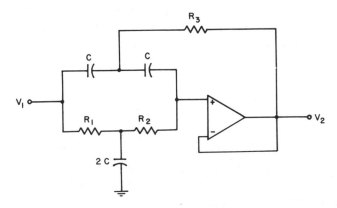

Figure 11.18. A second-order *VCVS* band-reject filter.

rad/s and $B = 1/Q$ rad/s, and take $C = 1$. Then the transfer function of Fig. 11.18 is given by

$$\frac{V_2}{V_1} = \frac{s^2 + G_1G_2}{s^2 + 2G_2s + G_1G_2} \tag{11.70}$$

provided

$$G_3 = G_1 + G_2 \tag{11.71}$$

Thus (11.69) is realized in the normalized case with a gain of $\dot{G} = 1$ and

$$\frac{1}{Q} = 2G_2$$
$$1 = G_1G_2 \tag{11.72}$$

Solving for the normalized conductances we have

$$G_1 = 2Q$$
$$G_2 = \frac{1}{2Q} \tag{11.73}$$
$$G_3 = 2Q + \frac{1}{2Q}$$

The denormalization process is carried out as before.

As an example, suppose we want a band-reject filter with the notch at $\omega_0 = 10^4$ rad/s, $Q = 5$, and a gain of 1, using capacitances of 0.01 μF. The normalized conductances are $G_1 = 10$, $G_2 = 0.1$, and $G_3 = 10.1$ mhos. The impedance scale factor is $k_i = 1/(10^4)(10^{-8}) = 10^4$, so that the denormalized resistances are $R_1 = 1k\Omega$, $R_2 = 100k\Omega$, and $R_3 = 990\Omega$.

Some advantages of the *VCVS* circuit are that it requires a minimal number of elements and has a noninverting gain. It may be seen from (11.73) that a high Q, say $Q > 10$, results in a wide spread of element values, which is not desirable. Thus Q should be restricted to values of 10 or less for best performance. From (11.72) we see that may adjust Q by varying R_2 and subsequently adjust the center frequency by varying R_1. A disadvantage of the circuit is that the gain is restricted to 1.

11.16 AN INFINITE-GAIN *MFB* BAND-REJECT FILTER

The circuit of Fig. 11.18, considered in the previous section, is limited to a gain of 1 and for best performance should not be used for very high Q. A band-reject circuit for which the gain may be specified and which is capable of achieving somewhat higher values of Q is shown in Fig. 11.19 [Hu-2]. Although the circuit is not exactly of the form of Fig. 10.14, since it has a second op amp and three additional resistors, we shall call it an infinite-gain *MFB* filter because of the multiple-feedback paths and the infinite gain mode of operation of the op amps.

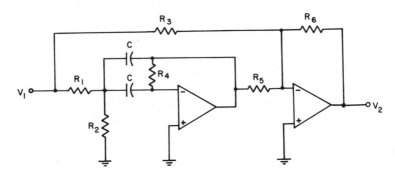

Figure 11.19. A second-order infinite-gain *MFB* band-reject filter.

Analysis of Fig. 11.19 shows it to achieve (11.69) with an inverting gain of

$$G = \frac{G_3}{G_6} \tag{11.74}$$

and

$$B = \frac{2G_4}{C}$$
$$\omega_0^2 = \frac{G_4(G_1 + G_2)}{C^2} \tag{11.75}$$

provided

$$G_1 G_5 = 2G_3 G_4 \tag{11.76}$$

For the normalized case, $\omega_0 = 1$ rad/s and $B = 1/Q$ rad/s, we select $C = 1F$, and thus two of the conductances may be assigned arbitrarily. There are many solutions of (11.74), (11.75), and (11.76). The one we choose, which gives relatively good values, is given by the normalized conductances,

$$G_1 = \frac{2}{Q}$$
$$G_2 = 2(Q - 1/Q)$$
$$G_3 = 1$$
$$G_4 = \frac{1}{2Q} \qquad (11.77)$$
$$G_5 = 0.5$$
$$G_6 = \frac{1}{G}$$

11.17 A BIQUAD BAND-REJECT FILTER

As in the bandpass case, a biquad band-reject circuit, though requiring more elements, is capable of much higher Qs, say up to 100, and is a much more stable circuit. The biquad we choose is that obtained from Fig. 10.26 which realizes (10.46) in the band-reject case of $d = 0$ and $e = bc = G$. Also we have in the normalized case,

$$C_1 = C_2 = G_7 = G_8 = G_9 = 1 \qquad (11.78)$$

and the values $b = 1$, $a = 1/Q$, and $c = G$.

For this case the biquad is shown in Fig. 11.20, where from (10.57) we have the normalized conductances

$$G_1 = \frac{G}{Q}$$
$$G_2 = \frac{1}{Q}$$
$$G_3 = 1 \qquad (11.79)$$
$$G_4 = G_5 = G$$

We have taken $G_6 = 0$, eliminating resistor R_6. The transfer function (11.69) is realized with inverting gain G by denormalization.

Eqs. (11.79) may be checked by noting that the transfer function of

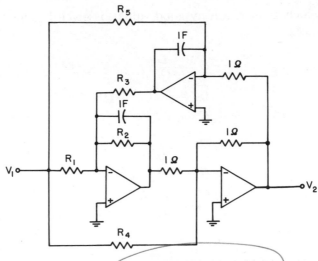

Figure 11.20. A biquad band-reject filter.

Fig. 11.20 is given by

$$\frac{V_2}{V_1} = -\frac{G_4[s^2 + (G_2 - G_1/G_4)s + G_3G_5/G_4]}{s^2 + G_2s + G_3} \qquad (11.80)$$

This result may also be used to obtain higher-order band-reject filters by cascading two or more sections, each of which must be matched to (11.80) for

$$G_2 = \frac{G_1}{G_4} \qquad (11.81)$$

For example, the response shown earlier in Fig. 4.9 is that of a fourth-order 1 dB Chebyshev band-reject filter with $Q = 10$, obtained by cascading two biquads.

11.18 A MULTIPLE-FEEDBACK ALL-PASS FILTER

The circuit of Fig. 11.21 due to Deliyannis [Del] realizes the second-order all-pass function

$$\frac{V_2}{V_1} = G\frac{s^2 - as + b}{s^2 + as + b} \qquad (11.82)$$

as may be easily shown, if

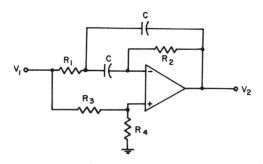

Figure 11.21. A multiple-feedback all-pass filter.

$$a = \frac{2G_2}{C}$$

$$b = \frac{G_1 G_2}{C^2} \tag{11.83}$$

$$G = \frac{G_3}{(G_3 + G_4)}$$

provided

$$4G_2 G_3 = G_1 G_4 \tag{11.84}$$

We shall refer to the circuit as a multiple-feedback circuit because of its construction, and to distinguish it from the biquad to be considered in the next section.

We note from (11.83) and (11.84) that for a given value of C, one of the conductances is arbitrary, and that $G < 1$. However, for minimum *dc* offset we have

$$R_2 = \frac{R_3 R_4}{(R_3 + R_4)}$$

which, along with (11.83) and (11.84), may be used to find the resistances,

$$R_1 = \frac{a}{2bC}$$

$$R_2 = \frac{2}{aC}$$

$$R_3 = \frac{2(a^2 + b)}{abC} \tag{11.85}$$

$$R_4 = \frac{2(a^2 + b)}{a^3 C}$$

The gain in this case is given by

$$G = \frac{b}{(a^2 + b)} \tag{11.86}$$

which is less than 1, since $b > 0$.

We may use (11.85) and (11.86) for the normalized case (a and b relatively simple numbers and C usually taken as $1F$) and denormalize the result for a given phase shift and damping constant at ω_0 rad/s (see Sec. 7.3). Or we may directly substitute the denormalized values of a and b and the desired value of C ($a = \omega_0 a'$ and $b = \omega_0^2 b'$, where a' and b' are the normalized values). In this case we have the denormalized resistances without scaling.

11.19 A BIQUAD ALL-PASS FILTER

We may realize (11.82) with a biquad all-pass section, obtained from Fig. 10.26. From (10.46) we see that we must have $c = G$, $d = -aG$, and $e = bG$. We shall consider the normalized network shown in Fig. 11.22, where

$$C_1 = C_2 = G_7 = G_8 = G_9 = 1 \tag{11.87}$$

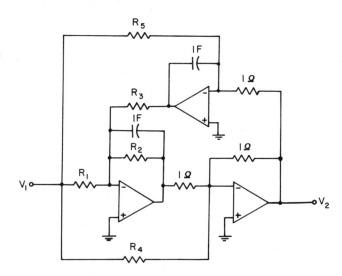

Figure 11.22. A biquad all-pass filter.

and by (10.57), choosing $G_6 = 0$, we have

$$G_1 = 2aG$$
$$G_2 = a \qquad\qquad\qquad (11.88)$$
$$G_3 = b$$
$$G_4 = G_5 = G$$

11.20 BESSEL FILTERS

Since the Bessel, or constant-time-delay filters are low-pass filters, we may synthesize them actively by the methods of Secs. 11.2 through 11.5. The only difference, aside from the low-pass coefficients, is that the frequency scale factor ω_0 is chosen in the manner described in Sec. 7.6 to yield a given time delay τ. The time delay will then be approximately τ over the interval $0 < \omega < \omega_0$.

For example, if we wish a second-order Bessel filter with $\tau = 0.1$ μs, we have from (7.43) and (7.49),

$$10^{-7} = \frac{12}{13\omega_0}$$

or $\omega_0 = 9.231 \times 10^6$ rad/s. We then realize the normalized second-order Bessel function

$$\frac{V_2}{V_1} = \frac{K}{s^2 + 3s + 3}$$

and denormalize the result for given capacitances and ω_0.

In the case of the all-pass, constant-time-delay filter of Sec. 7.8, we use the results of Secs. 11.18 and 11.19, obtaining ω_0 as in the Bessel filter case.

11.21 SUMMARY

In this chapter a number of active realizations are given for low-pass, high-pass, bandpass, band-reject, all-pass, and Bessel filters. The filters obtained are examples of the general networks of Chapter 10, as well as other standard, well-known circuits and some relatively new circuits.

Some advantages and disadvantages of each filter type are given, such as minimal number of elements required, relative ease of adjustment of characteristics, low output impedance, and stability.

A final concept which may determine whether one filter design has an advantage over another is that of sensitivity, a discusion of which has been left for Chapter 12. It might be argued that sensitivity should be given earlier and used in Chapter 11 in listing advantages and disadvantages. We have chosen to leave it until last, however, so that it will not be a prerequisite for reading the first eleven chapters. Thus in a short course in which one subject or another must be omitted, the reader has the study of sensitivity as an option.

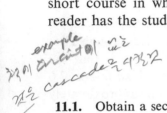

EXERCISES

11.1. Obtain a second-order infinite-gain *MFB* low-pass 1 dB Chebyshev filter with a gain of 2 and a cutoff frequency of 1,000 Hz, using equal capacitors of 0.01 μF.

11.2. Repeat Exercise 11.1 using a *VCVS* network.

11.3. Show that $G = 1$ is not possible if $C_1 = 1F$ in (11.8) for the Butterworth, 0.1, 0.5, 1, 2, and 3 dB Chebyshev cases.

11.4. Obtain a fourth-order low-pass Butterworth filter with a gain of 4 and a cutoff frequency of 1,000 Hz, using two cascaded *VCVS* sections having equal capacitors of 0.01 μF.

11.5. Repeat Exercises 11.2 and 11.4 using a biquad circuit.

11.6. Obtain a fourth-order low-pass filter using the state-variable method of Sec. 11.5. Using the result, obtain a fourth-order Butterworth filter with $G = 2$, $f_c = 1,000$ Hz, and all capacitors 0.01 μF. Compare the result with that of Exercise 11.4.

11.7. Show that in Fig. 11.6 for n even, if the output is taken at the input node of the inverter in series with $R_{n/2}$, then the result is given by

$$\frac{V_2}{V_1} = \frac{G_0(-s)^{n/2}}{s^n + G_1 s^{n-1} + \cdots + G_{n-1}s + G_n}; \qquad n = 2, 4, 6, \ldots$$

Thus the circuit is a bandpass filter of order n.

11.8. Obtain an active realization of the elliptic filter function of Exercise 6.10, using Fig. 11.8. Let $G = 2$, $f_c = 2,000$ Hz (end of the ripple channel), and use 0.01 μF capacitors.

11.9. Show that the elliptic function of (11.17) may be realized with non-inverting gain by the circuit shown in Fig. Ex. 11.9, provided $b_0 + 1 > b_1$, by obtaining $R_1 = b_0$, $R_2 = (b_0 + 1 - b_1)/b_1$, $R_3 = a_1$, $K = \mu(b_0 + 1 - b_1)/(a_1 + 1)$, gain $= G = Ka_1/b_0$, and $\mu = 1 + R_5/R_4$.

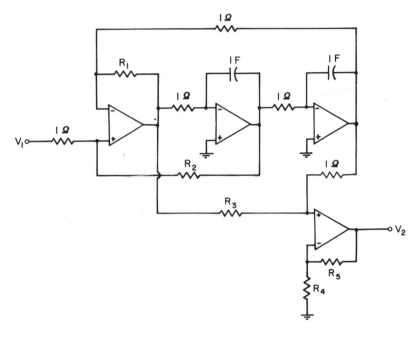

Figure Ex. 11.9

11.10. Repeat Exercise 11.1 for a high-pass filter.

11.11. Repeat Exercise 11.2 for a high-pass filter.

11.12. Obtain a fourth-order 0.5 dB Chebyshev high-pass filter with a gain of 4 and a cutoff frequency of 1,000 Hz, using two cascaded *VCVS* sections with all capacitances 0.01 μF.

11.13. Obtain a second-order 2 dB Chebyshev high-pass filter with a gain of 6 and a cutoff frequency of 10,000 Hz, using a biquad circuit with 0.1 μF capacitors.

11.14. Repeat Exercise 11.12 using biquad sections.

11.15. Obtain a biquad realization of the high-pass elliptic function whose low-pass prototype is given in Exercise 11.8.

11.16. A second-order bandpass filter is to have $f_0 = 1,000$ Hz, $Q = 10$, and $G = 5$, and the capacitors available are 0.01 μF. Realize these specifications with an infinite-gain *MFB* circuit.

11.17. Repeat Exercise 11.16 with a *VCVS* circuit.

11.18. Repeat Exercise 11.16 with a positive feedback circuit.

11.19. Repeat Exercise 11.16 with a biquad circuit.

11.20. Obtain a fourth-order Butterworth bandpass filter with $G = 25$, $f_0 = 1,000$ Hz, and $Q = 10$ by using only 0.01 μF capacitors in a cascade connection of two positive feedback networks.

11.21. Repeat Exercise 11.20 with two biquads.

11.22. Repeat Exercise 11.20 using the multiple-resonator network with biquad resonators.

11.23. Obtain an eighth-order 0.5 dB Chebyshev bandpass filter with $f_0 = 1,000$ Hz, $Q = 5$, and $G = 10$, using the multiple-resonator circuit with positive feedback resonators.

11.24. Verify (11.65) by substitution into (11.64).

11.25. A band-reject filter of order 2 is to have a center frequency of $f_0 = 60$ Hz and $Q = 10$. Using the *VCVS* network with $C = 0.1$ μF, obtain a realization for $G = 1$.

11.26. Repeat Exercise 11.25 with $G = 10$, using an infinite-gain *MFB* network.

11.27. Repeat Exercise 11.25 with $G = 10$, using a biquad.

11.28. Obtain a fourth-order 0.1 dB Chebyshev band-reject filter with $f_0 = 1,000$ Hz, $G = 25$, and $Q = 10$ using two biquads in cascade, with only 0.01 μF capacitors.

11.29. Obtain a *VCVS* realization of the all-pass, filter described in Exercise 7.1. Use only 0.01 μF capacitors.

11.30. Repeat Exercise 11.29 using a biquad circuit.

11.31. A Bessel filter of second-order is to have a constant time delay of approximately 1 μs over $0 < \omega < \omega_0$, and a gain of 2. Determine ω_0 and obtain a *VCVS* design using 0.01 μF capacitors.

11.32. Obtain a third-order Bessel filter having the specifications of Exercise 11.31, using a state variable circuit of Fig. 11.6.

11.33. Obtain a fourth-order Bessel filter having the specifications of Exercise 11.31, using two biquads in cascade.

11.34. Obtain a second-order all-pass, constant-time-delay filter, described in Sec. 7.8, by means of a biquad with capacitors of 0.01 μF, so that the time delay at $\omega_0 = 10{,}000$ rad/s is approximately 100 μs.

11.35. The circuit shown in Fig. Ex. 11.35 is an all-pass section [T-3]. Prove this by showing that it achieves (11.82) with $a = G_1/C$, $b = G_2 G_3/C^2$, $G = -G_6/G_5$, where $G_4 = 2G_1$.

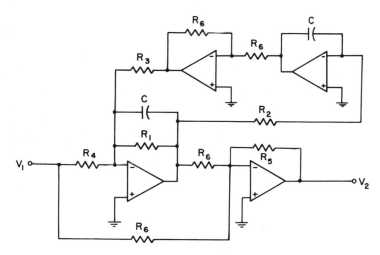

Figure Ex. 11.35.

CHAPTER

12

Sensitivity

12.1 DEFINITION

The characteristics of active and passive elements used in a filter design may vary from their nominal values because of aging, environmental changes, and other causes. These variations may cause a network to depart significantly from its desired performance. For example, in an active filter the gain of a $VCVS$ may change to the extent that transfer function poles are shifted to the right-half plane, resulting in instability. Thus a problem of interest in network theory is the *sensitivity* of the network characteristics to changes in the network parameters. The sensitivity problem is not serious for most resistively-terminated LC ladder networks [O], but is very important in active synthesis.

Formally, we define the measure of the change Δy in some performance characteristic y, resulting from a change Δx in a network parameter x to be the *sensitivity of y with respect to x*, given by

$$S_x^y = \frac{\Delta y/y}{\Delta x/x} = \frac{x}{y} \frac{\Delta y}{\Delta x} \tag{12.1}$$

Thus the changes in x and y have been *normalized*. That is, Δy is divided by y and Δx by x so that S_x^y is a ratio of normalized changes or percentages. For example, if $S_x^y = 0.5$, then a 2% change in x will cause a 1% change in y.

Eq. (12.1) may be put in a more useful form by considering the Taylor's series,

$$y + \Delta y = y + \frac{\partial y}{\partial x_1}\Delta x_1 + \frac{\partial y}{\partial x_2}\Delta x_2 + \cdots + \frac{\partial y}{\partial x_n}\Delta x_n$$

$$+ \text{ higher-order terms}$$

where y is a function of $x_1, x_2, \ldots, x_n$. Neglecting the higher order terms and letting x_i vary while the other $x_j, j \neq i$, are fixed, we have

$$\Delta y \approx \frac{\partial y}{\partial x_i}\Delta x_i$$

so that by (12.1), $S_{x_i}^y$ is approximately

$$S_{x_i}^y = \frac{x_i}{y}\frac{\partial y}{\partial x_i}, \qquad i = 1, 2, \ldots, n \qquad (12.2)$$

Theoretically (12.2) is only valid for small changes, but as a practical matter the sensitivity function is adequate for changes of network parameters up to 5 and sometimes 10 percent [GTH].

As an example, consider the overall transfer function H of n cascaded subnetworks,

$$H = \prod_{i=1}^{n} H_i \qquad (12.3)$$

which was considered earlier in Sec. 10.9. The functions H_i are the transfer functions of the subnetworks. The sensitivity of H with respect to a subnetwork function H_j is given by

$$S_{H_j}^H = \frac{H_j}{H}\frac{\partial H}{\partial H_j} = \frac{H_j}{\prod\limits_{i=1}^{n} H_i} \prod\limits_{\substack{i=1 \\ i \neq j}}^{n} H_i = 1$$

Thus a relative change in H_j results in the same relative change in H.

12.2 SOME SENSITIVITY IDENTITIES

From the definition of the sensitivity function S_x^y, given by

$$S_x^y = \frac{x}{y}\frac{\partial y}{\partial x} \qquad (12.4)$$

Sensitivity S_x^y can be used to determine the
$\%$ change in y due to a $\%$ change in x
(i.e) $\frac{\partial y}{y} = S_x^y \frac{\partial x}{x}$

OK, we can determine the absolute change in y due to an absolute change in X
by $\Delta y = \frac{y}{x} S_x^y \Delta x = \frac{\partial y}{\partial x} \Delta x$

(SEC. 12.2)   SOME SENSITIVITY IDENTITIES   **265**

we may derive a number of useful results. For example, if $y = u/v$, where u and v are functions of x, then we have

$$S_x^{u/v} = \frac{x}{u/v}\left[\frac{v\frac{\partial u}{\partial x} - u\frac{\partial v}{\partial x}}{v^2}\right]$$

which may be simplified, using (12.4), to

$$S_x^{u/v} = S_x^u - S_x^v \tag{12.5}$$

In a similar manner we have

$$S_x^{y^k} = \frac{x}{y^k}ky^{k-1}\frac{\partial y}{\partial x}$$

or

$$S_x^{y^k} = kS_x^y \tag{12.6}$$

where k is independent of x. If $k = -1$ in (12.6) [or $u = 1$ in (12.5)] we have

$$S_x^{1/y} = -S_x^y \tag{12.7}$$

Replacing v by $1/v$ in (12.5) and using (12.7) results in

$$S_u^{uv} = S_x^u + S_x^v \tag{12.8}$$

Suppose now we replace x by $1/x$ in (12.4). Then since for $z = 1/x$ we have

$$\frac{\partial y}{\partial z} = \frac{\partial y}{\partial x}\left[\frac{-1}{z^2}\right] = -x^2\frac{\partial y}{\partial x}$$

then (12.4) becomes

$$S_{1/x}^y = \frac{1}{xy}\left[-x^2\frac{\partial y}{\partial x}\right]$$

or

$$S_{1/x}^y = -S_x^y \tag{12.9}$$

Thus if $x = G_i$, a conductance corresponding to a resistance $R_i = 1/G_i$, we have

$$S_{R_i}^y = -S_{G_i}^y \tag{12.10}$$

Other results which may be obtained from the definition of sensitivity are

$$S_x^{ku} = S_x^u; \qquad k \text{ independent of } x \tag{12.11}$$

$$S_x^{e^u} = u S_x^u \tag{12.12}$$

$$S_x^{y(z_1, z_2, \ldots, z_n)} = \sum_{i=1}^{n} S_{z_i}^y S_x^{z_i}; \ z_i = z_i(x) \tag{12.13}$$

$$S_x^{\pi u_i} = \sum_i S_x^{u_i} \tag{12.14}$$

This last result is a generalization of (12.8), of course.

12.3 SOME SENSITIVITY FUNCTIONS

Let us consider the transfer function

$$H(s) = \frac{N(s)}{D(s)} \tag{12.15}$$

or for $s = j\omega$,

$$H(j\omega) = |H(j\omega)| e^{j\phi(\omega)} \tag{12.16}$$

By (12.5) and (12.15) we have the transfer function sensitivity,

$$S_x^H = S_x^N - S_x^D \tag{12.17}$$

or by (12.8) and (12.16) we have

$$S_x^{H(j\omega)} = S_x^{|H(j\omega)|} + S_x^{e^{j\phi(\omega)}}$$

which by (12.11) and (12.12) becomes

$$S_x^{H(j\omega)} = S_x^{|H(j\omega)|} + j\phi(\omega) S_x^{\phi(\omega)} \tag{12.18}$$

If x is real, then $S_x^{|H|}$ and S_x^{ϕ} are real and by (12.18) we have

$$S_x^{|H(j\omega)|} = \text{Re } S_x^{H(j\omega)} \tag{12.19}$$

and

$$S_x^{\phi(\omega)} = \frac{1}{\phi(\omega)} \text{Im } S_x^{H(j\omega)} \tag{12.20}$$

Thus the amplitude and phase sensitivities may be calculated directly by the definition, or they may be obtained from S_x^H using (12.19) and (12.20).

Let us now consider a general second-degree polynomial $D(s)$, which we write in the form

$$
\begin{aligned}
D(s) &= s^2 + a_1 s + a_0 \\
     &= (s - \sigma_1 - j\omega_1)(s - \sigma_1 + j\omega_1) \\
     &= s^2 + \frac{\omega_0}{Q} s + \omega_0^2
\end{aligned}
\tag{12.21}
$$

If $D(s)$ is the denominator of a second-order bandpass or band-reject filter function, then ω_0 and Q are respectively the center frequency and quality factor. In other cases, such as in low-pass or high-pass functions, or where $D(s)$ is a factor of a higher-degree denominator, we may still think of these parameters as the *pole-pair* Q and *pole-frequency* ω_0, with sensitivities defined by (12.4). We may also consider the *coefficient* sensitivities $S_x^{a_1}$ and $S_x^{a_0}$ defined by (12.4), which are evidently related to the Q and ω_0 sensitivities. Since by (12.21) we have

$$
(12\text{-}4); \quad S_x^y = \frac{x}{y}\frac{\partial y}{\partial x}
$$

$$
\omega_0 = \sqrt{a_0}, \quad Q = \frac{\sqrt{a_0}}{a_1}
\tag{12.22}
$$

then by (12.6) we have

$$
(12\text{-}6); \quad S_x^{y^k} = k\,S_x^y \qquad k = \text{independent of } x
$$

$$
S_x^{\omega_0} = \frac{1}{2} S_x^{a_0}
\tag{12.23}
$$

and by (12.5) and (12.6) we have

$$
(12\text{-}5); \quad S_x^{\frac{u}{v}} = S_x^u - S_x^v
$$

$$
S_x^Q = \frac{1}{2} S_x^{a_0} - S_x^{a_1}
\tag{12.24}
$$

From (12.23) and (12.24) we may obtain

$$
S_x^{a_0} = 2 S_x^{\omega_0}, \quad S_x^{a_1} = S_x^{\omega_0} - S_x^Q
\tag{12.25}
$$

As an example, let us consider the transfer function

$$
H(s) = \frac{\dfrac{G\omega_0}{Q} s}{s^2 + \dfrac{\omega_0}{Q} s + \omega_0^2}
\tag{12.26}
$$

which is that of a second-order bandpass filter with center frequency ω_0, quality factor Q, and gain G. Suppose we wish to find the sensitivity of the function $|H(j\omega)|$ to an element x. We note that

$$|H(j\omega)|^2 = \frac{G^2\omega_0^2\omega^2/Q^2}{(\omega_0^2 - \omega^2)^2 + \omega_0^2\omega^2/Q^2} = \frac{|N|^2}{|D|^2} \qquad (12.27)$$

so that, by using the identities developed earlier, we have

$$S_x^{|H|} = \frac{1}{2}S_x^{|H|^2} = \frac{1}{2}[S_x^{|N|^2} - S_x^{|D|^2}]$$

$$= \frac{1}{2}[S_x^{G^2} + S_x^{\omega_0^2} - S_x^{Q^2} - S_x^{|D|^2}]$$

$$= S_x^G + S_x^{\omega_0} - S_x^Q - \frac{1}{2}S_x^{|D|^2} \qquad (12.28)$$

From (12.27) we have

$$|D|^2 = (\omega_0^2 - \omega^2)^2 + \frac{\omega_0^2\omega^2}{Q^2} \qquad (12.29)$$

so that

$$S_x^{|D|^2} = \frac{x}{|D|^2}\left[2(\omega_0^2 - \omega^2)(2\omega_0)\frac{\partial\omega_0}{\partial x} + \frac{\omega^2}{Q^4}\left(2Q^2\omega_0\frac{\partial\omega_0}{\partial x} - 2Q\omega_0^2\frac{\partial Q}{\partial x}\right)\right]$$

This may be simplified to

$$S_x^{|D|^2}(\omega) = \frac{2\omega_0^2}{|D|^2}\left[2(\omega_0^2 - \omega^2)S_x^{\omega_0} + \frac{\omega^2}{Q^2}(S_x^{\omega_0} - S_x^Q)\right] \qquad (12.30)$$

where we have indicated that the sensitivity depends on ω. (The dependence on parameters such as x, as well as on ω_0 and Q, is tacitly understood.) Thus we may write (12.28) in the form

$$S_x^{|H|}(\omega) = S_x^G + S_x^{\omega_0} - S_x^Q - \frac{\omega_0^2}{|D|^2}\left[2(\omega_0^2 - \omega^2)S_x^{\omega_0} + \frac{\omega^2}{Q^2}(S_x^{\omega_0} - S_x^Q)\right] \qquad (12.31)$$

If the gain is given by

$$G = \frac{Q}{\omega_0} \qquad (12.32)$$

(in which case the numerator of $H(s)$ in (12.26) becomes s), then we have

$$S_x^G = S_x^Q - S_x^{\omega_0} \tag{12.32}$$

and (12.31) reduces to

$$S_x^{|H|}(\omega) = \frac{\omega_0^2}{|D|^2}\left[2(\omega^2 - \omega_0^2)S_x^{\omega_0} + \frac{\omega^2}{Q^2}(S_x^Q - S_x^{\omega_0})\right] \tag{12.34}$$

At $\omega = \omega_0$ we have $|D|^2 = \omega_0^4/Q^2$ from (12.29) and thus by (12.31) we have

$$S_x^{|H|}(\omega_0) = S_x^G + S_x^{\omega_0} - S_x^Q - S_x^{\omega_0} + S_x^Q$$
$$= S_x^G(\omega_0) \tag{12.35}$$

For G given by (12.32), we have from (12.33),

$$S_x^{|H|}(\omega_0) = S_x^Q(\omega_0) - S_x^{\omega_0}(\omega_0) \tag{12.36}$$

Let us now consider the behavior of $S_x^{|H|}$ for (12.26) at the cutoff points ω_1 and ω_2. As we recall, the cutoff points are related to Q and ω_0 by

$$\omega_1\omega_2 = \omega_0^2$$
$$\omega_2 - \omega_1 = B = \frac{\omega_0}{Q} \tag{12.37}$$

where B is the bandwidth. For $\omega = \omega_1$ we may write (12.31) in the form

$$S_x^{|H|}(\omega_1) = S_x^G + S_x^{\omega_0} - S_x^Q - \frac{\omega_1\omega_2\left[2(\omega_1\omega_2 - \omega_1^2)S_x^{\omega_0} + \frac{\omega_1^2}{Q^2}(S_x^{\omega_0} - S_x^Q)\right]}{(\omega_1\omega_2 - \omega_1^2)^2 + \omega_1^3\omega_2/Q^2}$$

or

$$S_x^{|H|}(\omega_1) = S_x^G + S_x^{\omega_0} - S_x^Q - \frac{\omega_2\{[2Q^2(\omega_2 - \omega_1) + \omega_1]S_x^{\omega_0} - \omega_1 S_x^Q\}}{Q^2(\omega_2 - \omega_1)^2 + \omega_1\omega_2}$$

Using (12.37) this may be simplified to

$$S_x^{|H|}(\omega_1) = S_x^G + S_x^{\omega_0} - S_x^Q + \frac{1}{2}(S_x^Q - S_x^{\omega_0}) - \frac{\omega_0}{\omega_1}QS_x^{\omega_0}$$

or

$$S_x^{|H|}(\omega_1) = S_x^G(\omega_1) + \frac{1}{2}[S_x^{\omega_0}(\omega_1) - S_x^Q(\omega_1)] - \frac{\omega_0}{\omega_1}QS_x^{\omega_0}(\omega_1) \tag{12.38}$$

If Q is large (B small), then $\omega_0 \approx \omega_1 + B/2$ and thus

$$\frac{\omega_0}{\omega_1} \approx 1 + \frac{B}{2\omega_1} \approx 1$$

so that

$$S_x^{|H|}(\omega_1) \approx S_x^G(\omega_1) + \left(\frac{1}{2} - Q\right)S_x^{\omega_0}(\omega_1) - \frac{1}{2}S_x^Q(\omega_1) \quad (12.39)$$

Similarly we may show that

$$S_x^{|H|}(\omega_2) = S_x^G(\omega_2) + \frac{1}{2}[S_x^{\omega_0}(\omega_2) - S_x^Q(\omega_2)] + \frac{\omega_0}{\omega_2}QS_x^{\omega_0}(\omega_2) \quad (12.40)$$

which for large Q becomes

$$S_x^{|H|}(\omega_2) \approx S_x^G(\omega_2) + \left(\frac{1}{2} + Q\right)S_x^{\omega_0}(\omega_2) - \frac{1}{2}S_x^Q(\omega_2) \quad (12.41)$$

12.4 · AN EXAMPLE

To illustrate the computation and use of sensitivity functions, let us consider the low-pass filter of Fig. 11.2, repeated as shown in Fig. 12.1, for the case $R_1 = 1/\sqrt{2}\,\Omega$, $R_2 = \sqrt{2}\,\Omega$, $C_1 = C_2 = 1F$, and the gain of the $VCVS$ given by x. We note that in this case the transfer function is given by (11.6) and (11.7) to be

$$H = \frac{V_2}{V_1} = \frac{x}{s^2 + [\sqrt{2} + (2 - x)/\sqrt{2}]s + 1} \quad (12.42)$$

Thus if the nominal value of gain is $x = 2$, (12.42) becomes the transfer function of a second-order low-pass Butterworth filter.

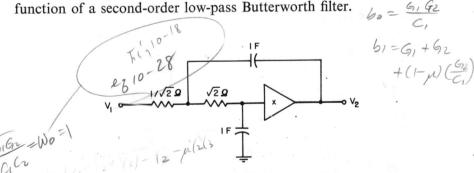

Figure 12.1. A second-order low-pass filter.

   In this example we are, of course, restricting our attention to changes in the network characteristics due to variations in x from its nominal value of $x = 2$. To consider other variations we would need to retain other parameters in the general case given by (11.6) and (11.7).

   Let us consider the sensitivity of $|H(j\omega)|$ with respect to the gain x, which we denote by $S_x^{|H|}(x, \omega)$, indicating its dependence on both the gain and the frequency. Using the results of the previous sections, we have

$$S_x^{|H|}(x, \omega) = \frac{1}{2} S_x^{|H|^2}(x, \omega)$$

$$= \frac{1}{2}[S_x^{x^2} - S_x^{|D|^2}] \tag{12.43}$$

where $|D|^2$ is the denominator of $|H|^2$, given by

$$|D|^2 = (1 - \omega^2)^2 + \frac{(4 - x)^2\omega^2}{2} \tag{12.44}$$

Computing the terms of (12.43) and simplifying yields

$$S_x^{|H|}(x, \omega) = 1 + \frac{x\omega^2(4 - x)}{2|D|^2} \tag{12.45}$$

For $x = 2$, (12.45) becomes

$$S_x^{|H|}(2, \omega) = 1 + \frac{2\omega^2}{1 + \omega^4} \tag{12.46}$$

from which we may show that $S_x^{|H|}(2, \omega)$ varies between 1 and 2, reaching its maximum value of 2 at the cutoff point $\omega_c = 1$ rad/s.

   Denoting $|H(j\omega)|$ by $|H(x, \omega)|$ to indicate its dependence on x, we may calculate the actual percentage change in $|H(x, \omega)|$ as x varies from 2, given by

$$\text{Actual change} = \frac{|H(x, \omega)| - |H(2, \omega)|}{|H(2, \omega)|} \times 100$$

This may be compared with the predicted percentage change, given by

$$\text{Predicted change} = S_x^{|H|}(2, \omega) \times \% \text{ change in } x$$

   Table 12.1 shows the results for various changes in x and various frequencies.

TABLE 12.1
Actual and Predicted Changes in $|H(x, \omega)|$

ω	1% increase in x		5% increase in x		10% increase in x	
	Actual	Predicted	Actual	Predicted	Actual	Predicted
0	1.0%	1.0%	5.0%	5.0%	10.0%	10.0%
1	2.05	2.0	10.54	10.0	22.2	20.0
2	1.48	1.47	7.51	7.36	15.29	14.71
10	1.0	1.02	5.0	5.1	10.0	10.2

For this example it is clear that although sensitivity functions are valid generally for small changes, they may be used with confidence for changes of 5 and even 10%. Of course, the smaller the change in x the better will be the agreement between actual and predicted changes in $|H(x, \omega)|$.

However, it should be noted that in this example, the sensitivity $S_x^{|H|}$ is low for x near its nominal value of 2. In this case the nominal value of Q is 0.707, which is low, as is typical of low-pass functions. On the other hand, if the sensitivity is high, the results in Table 12.1 will be quite different. For example, suppose $x = 3.9$, which results in a value of Q of 14.14 and a sensitivity at $\omega = 1$ of $S_x^{|H|}(3.9, 1) = 40$. Now a 1% increase in x yields a predicted change of 40% and an actual change of 65.58% in $|H(x, \omega)|$ for $\omega = 1$. We may conclude that the circuit of Fig. 12.1 is much more sensitive for high Q than for low Q, and that in very sensitive circuits, actual and predicted changes in network functions may be quite different even for small changes in element values.

From (11.6) and (11.7) we may calculate $S_x^{b_0}$ and $S_x^{b_1}$ for various parameters x and use (12.23) and (12.24) to obtain the Q and frequency sensitivities for the circuit of Fig. 12.1. The results are, for $C_2 = 1F$,

$$S_{G_1}^Q = \frac{1}{2} - \frac{G_1}{b_1} = -\frac{1}{2}, \qquad S_{G_1}^{\omega_0} = \frac{1}{2}$$

$$S_{G_2}^Q = -\frac{1}{2} + \frac{G_1}{b_1} = \frac{1}{2}, \qquad S_{G_2}^{\omega_0} = \frac{1}{2} \qquad (12.47)$$

$$S_{C_1}^Q = -\frac{1}{2} - \frac{(\mu - 1)G_2}{b_1 C_1} = -1, \qquad S_{C_1}^{\omega_0} = -\frac{1}{2}$$

where we have substituted the nominal values, $b_0 = 1$, $b_1 = \sqrt{2}$, $G_1 = \sqrt{2}$, $G_2 = 1/\sqrt{2}$, $C_1 = 1$, and $\mu = 2$. Also we have

$$S^\mu_{R_3} = -\frac{R_4}{\mu R_3} = -\frac{1}{2}, \qquad S^\mu_{R_4} = \frac{R_4}{\mu R_3} = \frac{1}{2} \qquad (12.48)$$

since for $\mu = 2$ we have $R_3 = R_4$. From these results we see that the circuit is not very sensitive to small changes in its element values. However, as we see from (12.47) before the nominal values are substituted, for a high gain μ, the pole-pair Q, and hence the location of the poles of the transfer function, could be very sensitive to changes in C_1.

12.5   ROOT SENSITIVITY

It is sometimes useful to define the sensitivity function without the normalization factor y of (12.1). Such an *unnormalized* function would then be given by

$$\mathcal{S}^y_x = x\frac{\partial y}{\partial x} \qquad (12.49)$$

Evidently this result is related to the sensitivity function of the previous sections by

$$\mathcal{S}^y_x = yS^y_x \qquad (12.50)$$

As an example of the use of $\mathcal{S}^y_x$, let us consider the function $F(\omega)$ expressed in dB by the relation

$$f(\omega) = 20 \log F(\omega) \qquad (12.51)$$

Then by (12.49) we have

$$\mathcal{S}^f_x = x\frac{\partial}{\partial x}20 \log F(\omega)$$

$$= x\frac{\partial}{\partial x}20 \log e \ln F(\omega)$$

$$= 8.686\frac{x}{F}\frac{\partial F}{\partial x}$$

$$= 8.686 \, S^F_x \qquad (12.52)$$

Thus the unnormalized sensitivity of f with respect to x is a factor times the sensitivity (in the classical sense) of F with respect to x.

The unnormalized sensitivity function is especially useful in considering changes in pole positions or *roots* of the transfer function denominator set equal to zero. For this reason, the unnormalized sensitivity is sometimes called *root sensitivity* [TM].

Suppose that $p_i = \sigma_i + j\omega_i$ is a pole of $H(s)$, or a zero of $D(s)$, as indicated for $i = 1$ in (12.21). Then by (12.49) we have

$$S_x^{p_i} = x\frac{\partial p_i}{\partial x} = x\frac{\partial}{\partial x}(\sigma_i + j\omega_i)$$

$$= S_x^{\sigma_i} + jS_x^{\omega_i} \tag{12.53}$$

If the denominator $D(s)$ is given by

$$\begin{aligned} D(s) &= s^2 + a_1 s + a_0 \\ &= (s - \sigma_1 - j\omega_1)(s - \sigma_1 + j\omega_1) \\ &= s^2 - 2\sigma_1 s + \sigma_1^2 + \omega_1^2 \end{aligned} \tag{12.54}$$

then we have

$$a_1 = -2\sigma_1, \qquad a_0 = \sigma_1^2 + \omega_1^2$$

or

$$\sigma_1 = -\frac{a_1}{2}, \qquad \omega_1^2 = a_0 - \frac{a_1^2}{4}$$

Thus we may obtain

$$S_x^{\sigma_1} = -\frac{x}{2}\frac{\partial a_1}{\partial x} = -\frac{a_1}{2}S_x^{a_1} \tag{12.55}$$

and

$$\begin{aligned} S_x^{\omega_1} &= x\frac{\partial}{\partial x}\left[a_0 - \frac{a_1^2}{4}\right]^{1/2} \\ &= \frac{x}{2\sqrt{a_0 - a_1^2/4}}\left[\frac{\partial a_0}{\partial x} - \frac{a_1}{2}\frac{\partial a_1}{\partial x}\right] \end{aligned}$$

or

$$S_x^{\omega_1} = \frac{1}{\sqrt{4a_0 - a_1^2}}\left[a_0 S_x^{a_0} - \frac{a_1^2}{2}S_x^{a_1}\right] \tag{12.56}$$

As an example, let us consider the transfer function (12.42), for which $a_0 = 1$ and a_1 is a function of the gain x, given by

$$\begin{aligned} a_1 &= \sqrt{2} + (2 - x)/\sqrt{2} \\ &= \sqrt{2}(2 - x/2) \end{aligned} \tag{12.57}$$

We have

$$S_x^{\sigma_1} = \frac{x}{2\sqrt{2}}, \qquad S_x^{\omega_1} = \frac{x(4-x)}{2\sqrt{16-2(4-x)^2}} \qquad (12.58)$$

which for the nominal value $x = 2$ becomes

$$S_x^{\sigma_1} = 1/\sqrt{2}, \qquad S_x^{\omega_1} = 1/\sqrt{2}$$

Therefore by (12.53) we have

$$S_x^{p_1} = 1/\sqrt{2} + j1/\sqrt{2} = 1\underline{|45°}$$

Since changes in p_1 due to changes in x are proportional to $S_x^{p_1}$, it follows that for x near 2, the root p_1 will change along a direction of 45°. For stability, a better design would be to choose x so that

$$\text{Arg } S_x^{p_1} = \arctan [S_x^{\omega_1}/S_x^{\sigma_1}]$$

given in (12.58), is near $\pm 90°$. Then p_1 would change in a direction parallel to the $j\omega$ axis and not tend to move toward the right-half plane. Of course, it is also desirable that $|S_x^{p_1}|$ be small.

12.6 VARIATIONS

Suppose the performance characteristic y is a function of the parameters $x_1, x_2, \ldots, x_n$. That is, $y = y(x_1, x_2, \ldots, x_n)$. Then we have

$$dy = \sum_{i=1}^{n} \frac{\partial y}{\partial x_i} dx_i = y \sum_{i=1}^{n} \left[\frac{x_i}{y} \frac{\partial y}{\partial x_i} \right] \frac{dx_i}{x_i}$$

or

$$\frac{dy}{y} = \sum_{i=1}^{n} S_{x_i}^y \frac{dx_i}{x_i} \qquad (12.59)$$

The quantity dy/y is the normalized change in y, and is sometimes called the *variation* of y. Thus we see that the variation of y is a weighted sum of the variations of the x_i. If the components x_i are *tracking* components (all x_i vary by the same percentage), then we have

$$\frac{dy}{y} = K \sum_{i=1}^{n} S_{x_i}^y$$

where $dx_i/x_i = K$, $i = 1, 2, \ldots, n$.

It is more practical, in an *RC*-active network, to assume that the resistances vary by one percentage and the capacitors by another percentage, so that the network has tracking components, by definition, if $dR_i/R_i = K_R$, $i = 1, 2, \ldots, N_R$, and $dC_j/C_j = K_C$, $j = 1, 2, \ldots, N_C$, where K_R and K_C are constants (not necessarily equal) and N_R and N_C are the number of resistances and capacitances, respectively. In this case we shall write

$$y = y(R_1, R_2, \ldots, R_{N_R}, C_1, C_2, \ldots, C_{N_C})$$

and (12.59) becomes

$$\frac{dy}{y} = \sum_{i=1}^{N_R} S_{R_i}^y \frac{dR_i}{R_i} + \sum_{j=1}^{N_C} S_{C_j}^y \frac{dC_j}{C_j} \tag{12.60}$$

In the case of tracking components, this becomes

$$\frac{dy}{y} = K_R \sum_{i=1}^{N_R} S_{R_i}^y + K_C \sum_{j=1}^{N_C} S_{C_j}^y \tag{12.61}$$

Suppose now that y is *homogeneous* in the R_i of *order* n_R, by which we mean that

$$y(\lambda R_1, \lambda R_2, \ldots, \lambda R_{N_R}, C_1, C_2, \ldots, C_{N_C})$$
$$= \lambda^{n_R} y(R_1, R_2, \ldots, R_{N_R}, C_1, C_2, \ldots, C_{N_C}) \tag{12.62}$$

Differentiating (12.62) with respect to λ and replacing λ by 1, we have

$$\sum_{i=1}^{N_R} \frac{\partial y}{\partial R_i} R_i = n_R y \tag{12.63}$$

By the same reasoning we may show that if y is homogeneous in the C_i of order n_C, that is,

$$y(R_1, R_2, \ldots, R_{N_R}, \lambda C_1, \lambda C_2, \ldots, \lambda C_{N_C})$$
$$= \lambda^{n_C} y(R_1, R_2, \ldots, R_{N_R}, C_1, C_2, \ldots, C_{N_C})$$

then

$$\sum_{j=1}^{N_C} \frac{\partial y}{\partial C_j} C_j = n_C y \tag{12.64}$$

Eqs. (12.63) and (12.64) may be written

$$\sum_{i=1}^{N_R} S_{R_i}^y = n_R \tag{12.65}$$

and

$$\sum_{j=1}^{N_C} S_{C_j}^y = n_C \tag{12.66}$$

so that if y is homogeneous in both the R_i and the C_j and the network has tracking components, (12.61) becomes

$$\frac{dy}{y} = K_R n_R + K_C n_C \tag{12.67}$$

Finally if, as is nearly always the case, y is a function of parameters $x_1, x_2, \ldots, x_M$, other than the R_i and C_j (gains of controlled sources, for example), then

$$y = y(R_1, \ldots, R_{N_R}, C_1, \ldots, C_{N_C}, x_1, \ldots, x_M)$$

and (12.60) becomes

$$\frac{dy}{y} = \sum_{i=1}^{N_R} S_{R_i}^y \frac{dR_i}{R_i} + \sum_{j=1}^{N_C} S_{C_j}^y \frac{dC_j}{C_j} + \sum_{k=1}^{M} S_{x_k}^y \frac{dx_k}{x_k} \tag{12.68}$$

If y is homogeneous in the R_i and C_j and the network has tracking R_i and C_j, then (12.68) may be written

$$\frac{dy}{y} = K_R n_R + K_C n_C + \sum_{k=1}^{M} S_{x_k}^y \frac{dx_k}{x_k} \tag{12.69}$$

As an example, let us consider the infinite-gain multiple-feedback bandpass filter of Fig. 10.17. Since $B = \omega_0/Q$ we may obtain from (10.26),

$$\omega_0 = \sqrt{\frac{1}{R_3 C_1 C_2}\left(\frac{1}{R_1} + \frac{1}{R_2}\right)} = f_1$$

$$Q = \frac{\sqrt{R_3 C_1 C_2\left(\frac{1}{R_1} + \frac{1}{R_2}\right)}}{C_1 + C_2} = f_2 \tag{12.70}$$

$$G = \frac{R_3 C_2}{R_1(C_1 + C_2)} = f_3$$

where $f_i = f_i(R_1, R_2, R_3, C_1, C_2)$, $i = 1, 2, 3$.

The function ω_0 is homogeneous of degree -1 in both the R_i and the C_i, since

$$f_1(\lambda R_1, \lambda R_2, \lambda R_3, C_1, C_2) = \sqrt{\frac{1}{\lambda R_3 C_1 C_2}\left(\frac{1}{\lambda R_1} + \frac{1}{\lambda R_2}\right)}$$
$$= \lambda^{-1} f_1(R_1, R_2, R_3, C_1, C_2)$$

and

$$f_1(R, R, R, \lambda C_1, \lambda C_2) = \sqrt{\frac{1}{R_3 \lambda C_1 \lambda C_2}\left(\frac{1}{R_1} + \frac{1}{R_2}\right)}$$
$$= \lambda^{-1} f_1(R_1, R_2, R_3, C_1, C_2)$$

By a similar procedure we see that Q and G are each homogeneous of degree 0 in both the R_i and the C_i. Thus if the components track, then by (12.67) we have

$$\frac{dQ}{Q} = \frac{dG}{G} = 0$$

and

$$\frac{d\omega_0}{\omega_0} = -K_R - K_C$$

12.7 SUMMARY

Sensitivity has been defined and a number of sensitivity functions and identities have been derived. These have been shown to be useful in determining the effect on the performance of the network due to deviations from nominal values of its elements. Needless to say, networks that are relatively insensitive are preferred over those that are highly sensitive to element changes. Also certain networks may be acceptable when used for one purpose but unacceptable when used for another purpose. (For example, a low-pass structure may perform well for low Q, as is typical in low-pass functions, but be too sensitive for high Q.)

This book has been intended as an introduction to filter theory and to a number of its various facets. Sensitivity is, of course, one of those facets and the treatment we have given is necessarily brief and of an introductory nature. For further study of this interesting subject, the reader is referred to Exercises 12.13–12.15, which consider third-order filters, and to the literature, which is rich and extensive.

EXERCISES

12.1. For the network shown in Fig. Ex. 12.1, find $H = V_2/V_1$, and obtain S_x^Q, $S_x^{\omega_0}$, and $S_x^{|H(j\omega)|}$ where x is successively each of the parameters, R, L, and C.

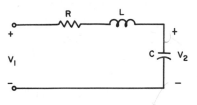

Figure Ex. 12.1.

12.2. Repeat Exercise 12.1 for the network shown in Fig. Ex. 12.2, where x is R_1, R_2, L, and C.

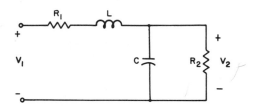

Figure Ex. 12.2.

12.3. Derive (12.11), (12.12), (12.13), and (12.14).

12.4. Find the sensitivities of Q, ω_0, and $|H(j\omega)|$ with respect to R_1, R_2, R_3, C_1, and C_2 for the network of Fig. 10.15. Evaluate the results for the nominal values $R_1 = R_2 = R_3 = 1\,\Omega$, $C_1 = \frac{1}{3}F$, and $C_2 = 1F$ (the Bessel filter).

12.5. Find the root sensitivity with respect to R_1 and R_2 for Exercise 12.4, evaluated at the given nominal values.

12.6. For Fig. 10.16, show that $R_2 = 1/R_1 = 2\sqrt{2}\,\Omega$, $C_1 = 2F$, and $C_2 = C_3 = 1F$ results in a Butterworth filter. Find the sensitivities of Q, ω_0, and $|H(j\omega)|$ with respect to each parameter, evaluated at the above nominal values.

12.7. Find S_x^Q and $S_x^{\omega_0}$ for Fig. 10.17, where x is R_1, R_2, R_3, C_1, and C_2.

12.8. Find the root sensitivity with respect to μ and R_1 for Fig. 10.19. Select normalized values of the elements, including $C_2 = 1F$, so as to yield a 1 dB Chebyshev filter with $\omega_c = 1$ rad/s and $G = 2$. (See (11.8) for the element values.) Evaluate the root sensitivities for these normalized values.

12.9. If the resistors all change by 5% and the capacitors all change by 10%, find the variations in Q, ω_0, and G for the circuit of Fig. 10.16, using the element values of Exercise 12.6.

12.10. Repeat Exercise 12.9 for the circuit of Fig. 10.21, assuming that μ changes by 5%. Use element values $C_1 = C_2 = 1F$, $\mu = 2$, and R_1 and R_2 so that the circuit is a Butterworth filter with $\omega_c = 1$ rad/s. Compare the results with those of Exercise 12.9.

12.11. In the low-pass biquad of Fig. 11.4, relabel the resistors of the inverter R_5 and R_6, where R_6 is connected to node a. Also relabel as C_1 the capacitor of the integrator. Show that

$$S_{R_1}^{\omega_0} = S_{R_2}^{\omega_0} = 0, \ S_{R_3}^{\omega_0} = S_{R_4}^{\omega_0} = S_{R_5}^{\omega_0} = -S_{R_6}^{\omega_0} = S_C^{\omega_0} = S_{C_1}^{\omega_0} = -1/2,$$

$$S_{R_1}^{Q} = 0, \ S_{R_2}^{Q} = 1, \ S_{R_3}^{Q} = S_{R_4}^{Q} = S_{R_5}^{Q} = -S_{R_6}^{Q} = -S_C^{Q} = S_{C_1}^{Q} = -1/2,$$

$$S_{R_2}^{G} = S_{R_4}^{G} = S_C^{G} = S_{C_1}^{G} = 0, \ S_{R_1}^{G} = -S_{R_3}^{G} = -S_{R_5}^{G} = S_{R_6}^{G} = -1.$$

Thus all the sensitivities S_i of the biquad are constants and satisfy the inequality, $0 \le |S_i| \le 1$, indicating that the circuit is extremely good from the standpoint of sensitivity.

12.12. (a) Given that

$$H(s) = \frac{H_1 H_2}{1 - K H_2}$$

where H_1 is a polynomial such as k, ks, or ks^2, and H_2 is the second-order transfer function of an RC network. Show that if the constant K affects only the coefficient of s in the denominator $s^2 + \dfrac{\omega_0}{Q} s + \omega_0^2$ of $H(s)$, then

$$S_K^Q > 2Q - 1$$

(b) Show that the Sallen and Key circuit of Fig. 11.2 is of this type, where $K = \mu$.
Suggestion: Note that

$$H_2 = \frac{N_1}{(s + a)(s + b)} \qquad 0 \le a < b$$

since it is an RC function, and we must have $N_1 = cs$ if K affects only the coefficient of s.

12.13. A number of second-order active filters are third-order filters reduced to second order by a pole-zero cancellation. (Consider, for example, the derivation of the transfer function (11.70) for Fig. 11.18.) Thus it is desirable to have a procedure for obtaining sensitivities of third-order systems. Let the denominator of the third-order transfer function by given by

$$D(s) = (s + a)\left(s^2 + \frac{\omega_0}{Q}s + \omega_0^2\right)$$

$$= s^3 + b_2 s^2 + b_1 s + b_0$$

and show that

$$b_0 S_x^{b_0} = a\omega_0^2(2S_x^{\omega_0} + S_x^a)$$

$$b_1 S_x^{b_1} = \omega_0\left[\left(2\omega_0 + \frac{a}{Q}\right)S_x^{\omega_0} - \frac{a}{Q}(S_x^Q - S_x^a)\right]$$

$$b_2 S_x^{b_2} = \frac{\omega_0}{Q}(S_x^{\omega_0} - S_x^Q) + aS_x^a$$

12.14. Solve the last three equations in Exercise 12.13 and thus show that the sensitivities of interest are given by

$$S_x^{\omega_0} = \frac{(aQ - \omega_0)b_0 S_x^{b_0} + \omega_0^2 Q b_1 S_x^{b_1} - a\omega_0^2 Q b_2 S_x^{b_2}}{2\omega_0^2(a^2 Q - a\omega_0 + \omega_0^2 Q)}$$

$$S_x^Q = \frac{(2\omega_0 Q^2 + aQ - \omega_0)b_0 S_x^{b_0} + \omega_0 Q(\omega_0 - 2aQ)b_1 S_x^{b_1} + \omega_0^2 Q(a - 2\omega_0 Q)b_2 S_x^{b_2}}{2\omega_0^2(a^2 Q - a\omega_0 + \omega_0^2 Q)}$$

$$S_x^a = \frac{Q(b_0 S_x^{b_0} - ab_1 S_x^{b_1} + a^2 b_2 S_x^{b_2})}{a(a^2 Q - a\omega_0 + \omega_0^2 Q)}$$

(The development in this and the previous exercise is based on [TM], pp. 342–343.)

12.15. Show that before pole-zero cancellation the transfer function for Fig. 11.18 is given by

$$\frac{V_2}{V_1} = \frac{s^2(2s + G_1 + G_2) + G_1 G_2(2s + G_3)}{(2s + G_1 + G_2)(s^2 + 2G_2 s + G_2 G_3) - G_3^2(2s + G_3)}$$

and if $G_1 = 16$, $G_2 = 1$, and $G_3 = 17$ mhos, then the nominal values of Exercise 12.13 are $a = 17/2$, $\omega_0 = 4$, and $Q = 2$. Using the results of Exercise 12.14, find $S_x^{\omega_0}$, S_x^Q, and S_x^a for $x = G_3$.

Bibliography

B   S. Butterworth, "On the theory of filter amplifiers," *Wireless Engineer*, vol. 7, pp. 536–541, October 1930.

Ba   N. Balabanian, *Network Synthesis*, Prentice-Hall, Inc., Englewood Cliffs, N.J., 1958.

Br   O. Brune, "Synthesis of a finite two-terminal network whose driving-point impedance is a prescribed function of frequency," *J. Math. Phys.*, vol. 10, pp. 191–236, August 1931.

Bu-1   A. Budak, "A maximally flat phase and controllable magnitude approximation," *IEEE Trans. on Circuit Theory*, vol. CT-12, p. 279, June 1965.

Bu-2   A. Budak, *Passive and Active Network Analysis and Synthesis*, Houghton Mifflin Co., Boston, 1974.

D   S. Darlington, "Synthesis of reactance 4-poles which produce prescribed insertion loss characteristics," *J. Math. Phys.*, vol. 18, pp. 257–353, September 1939.

De   C. A. Desoer, "Notes commenting on Darlington's design procedure for networks made of uniformly dissipative coils $(d_o + \delta)$ and uniformly dissipative capacitors $(d_o - \delta)$," *IRE Trans. on Circuit Theory*, vol. CT-6, no. 4, pp. 397–398, December 1959.

Del   T. Deliyannis, "RC active allpass sections," *Electron. Letters*, vol. 5, pp. 59–60, February 1969.

F   R. M. Foster, "A reactance theorem," *BSTJ*, vol. 3, pp. 259–267, April 1924.

FT   P. E. FLEISCHER and J. TOW, "Design formulas for biquad active filters using three operational amplifiers," *Proc. of the IEEE*, vol. 61, no. 5, pp. 662–663, May 1973.

G   E. A. GUILLEMIN, *Synthesis of Passive Networks*, John Wiley and Sons, Inc., New York, 1957.

Ge-1   P. R. GEFFE, *Simplified Modern Filter Design*, Hayden Book Co., Inc., New York, 1963.

Ge-2   P. R. GEFFE, "A note on predistortion," *IRE Trans. on Circuit Theory*, vol. CT-6, no. 4, p. 395, December 1959.

GTH   J. G. GRAEME, G. E. TOBEY, and L. P. HUELSMAN (eds.), *Operational Amplifiers: Design and Applications*, McGraw-Hill Book Co., New York, 1971.

H   D. S. HUMPHERYS, *The Analysis, Design, and Synthesis of Electrical Filters*, Prentice-Hall, Inc., Englewood Cliffs, N.J., 1970.

Ha   S. S. HAYKIN, *Active Network Theory*, Addison Wesley Publishing Co., Reading, Mass., 1970.

Hg   G. HURTIG III, "The primary resonator block technique of filter synthesis," *Proc. International Filter Symposium*, p. 84, April 1972.

HJ   J. L. HILBURN and D. E. JOHNSON, *Manual of Active Filter Design*, McGraw-Hill Book Co., New York, 1973.

HJE   J. L. HILBURN, D. E. JOHNSON, and A. ESKANDAR, "Low-pass and high-pass designs for higher-order active filters," *Proc. 1974 IEEE Region 3 Conference and Exhibit*, April 1974.

Hu-1   L. P. HUELSMAN, *Theory and Design of Active RC Circuits*, McGraw-Hill Book Co., New York, 1968.

Hu-2   L. P. HUELSMAN, *Active Filters: Lumped, Distributive, Integrated, Digital, and Parametric*, McGraw-Hill Book Co., New York, 1970.

I   R. M. INIGO, "Active filter realization using finite-gain voltage amplifiers," *IEEE Trans. on Circuit Theory*, vol. CT-17, pp. 445–448, August 1970.

JH   D. E. JOHNSON and J. L. HILBURN, *Rapid, Practical Design of Active Filters*, John Wiley and Sons, Inc., New York, 1975.

JHI   D. E. JOHNSON, J. L. HILBURN, and F. H. IRONS, "Multiple-feedback higher-order active filters," *Proc. 1974 IEEE Region 3 Conference and Exhibit*, April 1974.

JJ   D. E. JOHNSON and J. R. JOHNSON, *Mathematical Methods in Engineering and Physics*, The Ronald Press Publishing Co., New York, 1965.

JJK   D. E. JOHNSON, J. R. JOHNSON, and M. D. KASHEFI, "Ultraspherical rational filters," *IEEE Trans. on Circuit Theory*, vol. CT-20, pp. 596–599, September 1973.

K   F. F. KUO, *Network Analysis and Synthesis* (second edition), John Wiley and Sons, Inc., New York, 1966.

KF   H. H. KRALL and O. FRINK, "A new class of orthogonal polynomials: the Bessel polynomials," *Trans. Am. Math. Soc.*, vol. 65, pp. 100–115, January 1949.

KH   W. J. KERWIN and L. P. HUELSMAN, "The design of high performance active RC band-pass filters," *IEEE International Convention Record*, vol. 14, pt. 10, pp. 74–80, 1960.

KHN   W. J. KERWIN, L. P. HUELSMAN, and R. W. NEWCOMB, "State-variable synthesis for insensitive integrated circuit transfer functions," *IEEE Journal of Solid-State Circuits*, vol. SC-2, no. 3, pp. 87–92, September 1967.

L   W. F. LOVERING, "Analog computer simulation of transfer functions," *Proc. IEEE*, vol. 53, no. 3, pp. 306–307, March 1965.

M-1   S. K. MITRA, *Analysis and Synthesis of Linear Active Networks*, John Wiley and Sons, Inc., New York, 1969.

M-2   S. K. MITRA, "A network transformation for active RC networks," *Proc. IEEE*, vol. 55, no. 11, pp. 2021–2022, November 1967.

M-3   S. K. MITRA (ed.), *Active Inductorless Filters*, IEEE Press, New York, 1971.

MG   R. MELEN and H. GARLAND, *Understanding IC Operational Amplifiers*, Howard W. Sams and Co., New York, 1971.

MJJ   A. H. MARSHAK, D. E. JOHNSON, and J. R. JOHNSON, "A Bessel rational filter," *IEEE Trans. on Circuits and Systems*, vol. CAS-22, November 1974.

O   H. J. ORCHARD, "Inductorless filters," *Electron. Letters*, vol. 2, no. 6, pp. 224–225, June 1966.

P-1   A. PAPOULIS, "On the approximation problem in filter design," *IRE National Convention Record*, vol. 5, pt. 2, pp. 175–185, 1957.

P-2   A. PAPOULIS, *The Fourier Integral and its Applications*, McGraw-Hill Book Co., New York, 1962.

R   P. RICHARDS, "Universal optimum-response curve for arbitrarily selected coupled resonators," *Proc. IRE*, vol. 34, pp. 624–629, September 1946.

RB   H. RUSTON and J. BORDOGNA, *Electric Networks: Functions, Filters, Analysis*, McGraw-Hill Book Co., New York, 1966.

S-1   K. L. SU, *Time-Domain Synthesis of Linear Networks*, Prentice-Hall, Inc., Englewood Cliffs, N.J., 1971.

S-2   K. L. SU, *Active Network Synthesis*, McGraw-Hill Book Co., New York, 1965.

SK   R. P. SALLEN and E. L. KEY, "A practical method of designing RC active filters," *IRE Trans. on Circuit Theory*, vol. CT-2, pp. 74–85, March 1955.

St   L. STORCH, "Synthesis of constant-time-delay ladder networks using Bessel polynomials," *Proc. IRE*, vol. 42, no. 11, pp. 1666–1675, November 1954.

SU   R. SAAL and E. ULBRICH, "On the design of filters by synthesis," *IRE Trans. on Circuit Theory*, vol. CT-5, no. 4, pp. 284–327, December 1958.

Sz   G. SZENTIRMAI, "Synthesis of multiple-feedback active filters," *BSTJ*, vol. 52, no. 4, pp. 527–555, April 1973.

T-1   J. Tow, "Active RC filters—a state-space realization," *Proc. IEEE*, vol. 56, pp. 1137–1139, June 1968.

T-2   J. Tow, "Design formulas for active RC filters using operational-amplifier biquad," *Electron. Letters*, vol. 5, no. 15, pp. 339–341, 24 July 1969.

T-3   J. Tow, "A step-by-step active-filter design," *IEEE Spectrum*, vol. 6, no. 12, pp. 64–68, December 1969.

Th   W. E. THOMSON, "Delay networks having maximally flat frequency characteristics," *Proc. IEE*, pt. 3, vol. 96, pp. 487–490, November 1949.

TM   G. C. TEMES and S. K. MITRA, *Modern Filter Theory and Design*, John Wiley and Sons, Inc., New York, 1973.

Tu   D. F. TUTTLE, JR., *Network Synthesis Vol. I*, John Wiley and Sons, Inc., New York, 1958.

V   M. E. VAN VALKENBURG, *Introduction to Modern Network Synthesis*, John Wiley and Sons, Inc., New York, 1962.

W   L. WEINBERG, *Network Analysis and Synthesis*, McGraw-Hill Book Co., New York, 1962.

Appendices

Appendices

A

Low-Pass Coefficients

Coefficients of denominator polynomial
$$s^n + b_{n-1}s^{n-1} + \cdots + b_2s^2 + b_1s + b_0$$
of low-pass Butterworth and Chebyshev
transfer functions, normalized to a cutoff
frequency of 1 rad/s and of Bessel transfer
functions for $T(1)$ approximately 1 s

TABLE A.1
Butterworth Filter: $s^n + b_{n-1}s^{n-1} + \cdots + b_1s + b_0$

n	b_0	b_1	b_2	b_3	b_4	b_5	b_6	b_7
1	1.00000							
2	1.00000	1.41421						
3	1.00000	2.00000	2.00000					
4	1.00000	2.61313	3.41421	2.61313				
5	1.00000	3.23607	5.23607	5.23607	3.23607			
6	1.00000	3.86370	7.46410	9.14162	7.46410	3.86370		
7	1.00000	4.49396	10.09783	14.59179	14.59179	10.09783	4.49396	
8	1.00000	5.12583	13.13707	21.84615	25.68836	21.84615	13.13707	5.12583

TABLE A.2

0.1 dB Chebyshev Filter ($\epsilon = 0.15262$) : $s^n + b_{n-1}s^{n-1} + \cdots + b_1 s + b_0$

n	b_0	b_1	b_2	b_3	b_4	b_5	b_6	b_7
1	6.55222							
2	3.31329	2.37209						
3	1.63809	2.62953	1.93883					
4	0.82851	2.02550	2.62680	1.80377				
5	0.40951	1.43556	2.39696	2.77071	1.74396			
6	0.20713	0.90176	2.04784	2.77908	2.96575	1.71217		
7	0.10238	0.56179	1.48293	2.70514	3.16925	3.18350	1.69322	
8	0.05179	0.32645	1.06667	2.15932	3.41855	3.56485	3.41297	1.68104

TABLE A.3

0.5 dB Chebyshev Filter ($\epsilon = 0.34931$) : $s^n + b_{n-1}s^{n-1} + \cdots + b_1 s + b_0$

n	b_0	b_1	b_2	b_3	b_4	b_5	b_6	b_7
1	2.86278							
2	1.51620	1.42562						
3	0.71569	1.53490	1.25291					
4	0.37905	1.02546	1.71687	1.19739				
5	0.17892	0.75252	1.30957	1.93737	1.17249			
6	0.09476	0.43237	1.17186	1.58976	2.17184	1.15918		
7	0.04473	0.28207	0.75565	1.64790	1.86941	2.41265	1.15122	
8	0.02369	0.15254	0.57356	1.14859	2.18402	2.14922	2.65675	1.14608

TABLE A.4

1 dB Chebyshev Filter ($\epsilon = 0.50885$) : $s^n + b_{n-1}s^{n-1} + \cdots + b_1 s + b_0$

n	b_0	b_1	b_2	b_3	b_4	b_5	b_6	b_7
1	1.96523							
2	1.10251	1.09773						
3	0.49131	1.23841	0.98834					
4	0.27563	0.74262	1.45392	0.95281				
5	0.12283	0.58053	0.97440	1.68882	0.93682			
6	0.06891	0.30708	0.93935	1.20214	1.93083	0.92825		
7	0.03071	0.21367	0.54862	1.35754	1.42879	2.17608	0.92312	
8	0.01723	0.10734	0.44783	0.84682	1.83690	1.65516	2.42303	0.91981

TABLE A.5

2 dB Chebyshev Filter ($\epsilon = 0.76478$): $s^n + b_{n-1}s^{n-1} + \cdots + b_1 s + b_0$

n	b_0	b_1	b_2	b_3	b_4	b_5	b_6	b_7
1	1.30756							
2	0.82302	0.80382						
3	0.32689	1.02219	0.73782					
4	0.20577	0.51680	1.25648	0.71622				
5	0.08172	0.45935	0.69348	1.49954	0.70646			
6	0.05144	0.21027	0.77146	0.86701	1.74586	0.70123		
7	0.02042	0.16609	0.38251	1.14444	1.03922	1.99353	0.69789	
8	0.01286	0.07294	0.35870	0.59822	1.57958	1.21171	2.24225	0.69606

TABLE A.6

3 dB Chebyshev Filter ($\epsilon = 0.99763$): $s^n + b_{n-1}s^{n-1} + \cdots + b_1 s + b_0$

n	b_0	b_1	b_2	b_3	b_4	b_5	b_6	b_7
1	1.00238							
2	0.70795	0.64490						
3	0.25059	0.92835	0.59724					
4	0.17699	0.40477	1.16912	0.58158				
5	0.06264	0.40794	0.54886	1.41498	0.57443			
6	0.04425	0.16343	0.69910	0.69061	1.66285	0.57070		
7	0.01566	0.14615	0.30002	1.05184	0.83144	1.91155	0.56842	
8	0.01106	0.05648	0.32076	0.47190	1.46670	0.97195	2.16071	0.56695

TABLE A.7

Bessel Filter: $s^n + b_{n-1}s^{n-1} + \cdots + b_1 s + b_0$

n	b_0	b_1	b_2	b_3	b_4	b_5
1	1					
2	3	3				
3	15	15	6			
4	105	105	45	10		
5	945	945	420	105	15	
6	10,395	10,395	4,725	1,260	210	21

B

Low-Pass Second-Order Factors

Second-order factors of denominator polynomial

$$\prod_{i=1}^{n} (s^2 + a_i s + b_i)$$

of low-pass Butterworth and Chebyshev transfer functions, normalized to a cutoff frequency of 1 rad/s and of Bessel transfer functions for $T(1)$ approximately 1 s

TABLE B.1

Butterworth Filter: $\prod_{i=1}^{n} (s^2 + a_i s + b_i)$

n	a_1	b_1	a_2	b_2	a_3	b_3	a_4	b_4
1	1.41421	1.00000						
2	0.76537	1.00000	1.84776	1.00000				
3	0.51764	1.00000	1.41421	1.00000	1.93185	1.00000		
4	0.39018	1.00000	1.11114	1.00000	1.66294	1.00000	1.96157	1.00000

TABLE B.2

0.1 dB Chebyshev Filter: $\prod_{i=1}^{n} (s^2 + a_i s + b_i)$

n	a_1	b_1	a_2	b_2	a_3	b_3	a_4	b_4
1	2.37209	3.31329						
2	0.52827	1.32981	1.27536	0.62282				
3	0.22940	1.12953	0.62674	0.69646	0.85614	0.26339		
4	0.12797	1.06964	0.36443	0.79901	0.54540	0.41627	0.64334	0.14563

TABLE B.3

0.5 dB Chebyshev Filter: $\prod_{i=1}^{n} (s^2 + a_i s + b_i)$

n	a_1	b_1	a_2	b_2	a_3	b_3	a_4	b_4
1	1.42562	1.51620						
2	0.35071	1.06352	0.84668	0.35641				
3	0.15530	1.02302	0.42429	0.59001	0.57959	0.15700		
4	0.08724	1.01193	0.24844	0.74133	0.37182	0.35865	0.43859	0.08805

TABLE B.4

1 dB Chebyshev Filter: $\prod_{i=1}^{n} (s^2 + a_i s + b_i)$

n	a_1	b_1	a_2	b_2	a_3	b_3	a_4	b_4
1	1.09773	1.10251						
2	0.27907	0.98650	0.67374	0.27940				
3	0.12436	0.99073	0.33976	0.55772	0.46413	0.12471		
4	0.07002	0.99414	0.19939	0.72354	0.29841	0.34086	0.35200	0.07026

TABLE B.5

2 dB Chebyshev Filter: $\prod_{i=1}^{n} (s^2 + a_i s + b_i)$

n	a_1	b_1	a_2	b_2	a_3	b_3	a_4	b_4
1	0.80382	0.82325						
2	0.20977	0.92868	0.50644	0.22157				
3	0.09395	0.96595	0.25667	0.53294	0.35061	0.09993		
4	0.05298	0.98038	0.15089	0.70978	0.22582	0.32710	0.26637	0.05650

TABLE B.6

3 dB Chebyshev Filter: $\prod\limits_{i=1}^{n} (s^2 + a_i s + b_i)$

n	a_1	b_1	a_2	b_2	a_3	b_3	a_4	b_4
1	0.64490	0.70795						
2	0.17034	0.90309	0.41124	0.19598				
3	0.07646	0.95483	0.20890	0.52182	0.28535	0.08880		
4	0.04316	0.97417	0.12290	0.70358	0.18393	0.32089	0.21696	0.05029

TABLE B.7

Bessel Filter: $\prod\limits_{i=1}^{n} (s^2 + a_i s + b_i)$

n	a_1	b_1	a_2	b_2	a_3	b_3
1	3.00000	3.00000				
2	5.79242	9.14012	4.20758	11.48781		
3	8.49672	18.80114	7.47142	20.85282	5.03186	26.51399

C

Derivation
of the Positive-Real
Conditions

Let us consider the network N of Fig. C.1, which is assumed to be passive, with a given driving-point impedance $Z(s)$, assumed to be a rational function with real coefficients and a Hurwitz numerator and denominator.

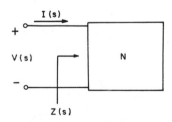

Figure C.1.

We know that the current response,

$$i(t) = \mathcal{L}^{-1}[I(s)] = \mathcal{L}^{-1}[V(s)/Z(s)]$$

has two components, $i(t) = i_{\text{free}} + i_{\text{forced}}$, where i_{free} is a function whose frequencies are the natural frequencies of the network, and i_{forced} has the characteristics of the driving function $v(t)$. The natural frequencies are the poles of the transfer function $1/Z(s)$, and consequently cannot be in the right-half of the s-plane. Therefore, if the network is excited by the voltage

$$v(t) = e^{\sigma_1 t} \cos \omega_1 t \qquad (C.1)$$

where $\sigma_1 > 0$, then the frequency of the driver,

$$s_1 = \sigma_1 + j\omega_1 = |s_1| e^{j\theta} \qquad (C.2)$$

cannot coincide with any of the natural frequencies, in which case $i_{\text{free}} \ll i_{\text{forced}}$ for t sufficiently large. Thus after a sufficient time has elapsed, we may ignore the free component of current and consider the current as approximately the forced component given by

$$i(t) = \frac{e^{\sigma_1 t} \cos(\omega_1 t - \phi)}{|Z|} \qquad (C.3)$$

where we define

$$Z(s_1) = |Z| e^{j\phi} \qquad (C.4)$$

The energy delivered to the network in t_1 seconds is given by

$$w(t_1) = \int_0^{t_1} vi \, dt = \int_0^{t_1} \frac{e^{2\sigma_1 t} \cos \omega_1 t \cos (\omega_1 t - \phi) \, dt}{|Z|}$$

which may be written

$$w = \frac{1}{2} \int_0^{t_1} \frac{e^{2\sigma_1 t}[\cos \phi + \cos(2\omega_1 t - \phi)] \, dt}{|Z|} \qquad (C.5)$$

Writing

$$e^{2\sigma_1 t} \cos (2\omega_1 t - \phi) = Re \; e^{2s_1 t - j\phi}$$

we may evaluate (C.5), obtaining

$$w = \frac{1}{4\sigma_1 |Z|}(e^{2\sigma_1 t} - 1) \cos \phi + Re\left\{\frac{e^{-j\phi}}{4|Z|s_1}(e^{2s_1 t_1} - 1)\right\}$$

This may be simplified, using $\sigma_1 = |s_1| \cos \theta$, to

$$w = \frac{1}{4|s_1||Z|}\left\{e^{2\sigma_1 t}\left[\frac{\cos \phi}{\cos \theta} + \cos (2\omega_1 t - \theta - \phi)\right]\right.$$
$$\left. - \cos(\theta + \phi) - \frac{\cos \phi}{\cos \theta}\right\} \qquad (C.6)$$

Since for a passive network w must be nonnegative, we see from (C.6) that it is necessary for the coefficient of $e^{2\sigma_1 t}$ to be nonnegative. That is,

$$\frac{\cos \phi}{\cos \theta} + \cos (2\omega_1 t_1 - \theta - \phi) \geq 0 \tag{C.7}$$

Since as t_1 increases, $\cos (2\omega_1 t_1 - \theta - \phi)$ varies between -1 and 1, (C.7) can be satisfied for all t_1 only if

$$\frac{\cos \phi}{\cos \theta} \geq 1$$

or since $|\theta| \leq \pi/2$, we have

$$\cos \phi \geq \cos \theta \tag{C.8}$$

Eq. (C.8) is equivalent to

$$|\phi| \leq |\theta| \tag{C.9}$$

and to

$$\frac{Re\ Z(s_1)}{|Z(s_1)|} \geq \frac{Re\ s_1}{|s_1|} \tag{C.10}$$

both of which are true for $Re\ s_1 > 0$. Therefore we may say that for $Re\ s > 0$, we have from (C.9)

$$|\operatorname{Arg} Z(s)| \leq |\operatorname{Arg} s| \tag{C.11}$$

and from (C.10)

$$Re\ Z(s) \geq 0 \tag{C.12}$$

Also from (C.11) we see that if s is real and positive, then $\operatorname{Arg} s = 0$, and hence $\operatorname{Arg} Z(s) = 0$, which is to say that when s is real and positive, then $Z(s)$ is real. Summing up, we have for $Re\ s > 0$,

$$\begin{aligned} &Z(s) \text{ is real when } s \text{ is real} \\ &Re\ Z(s) \geq 0 \end{aligned} \tag{C.13}$$

and also (C.11) holds.

Conditions (C.11) and (C.13) are not independent, as we have seen. Indeed, it is clear that (C.11) implies (C.13), although the converse is not evident. However, Brune [Br] showed that the converse was true and hence we may use the two conditions interchangeably. Functions which satisfy (C.13), and hence (C.11), were called *positive real* functions by Brune. Thus we have seen that if a one-port network is passive, then its driving-point impedance is positive real. The converse is also true, as Brune showed. That is, if $Z(s)$ is positive real, then there exists a passive realization.

Index